GUNNER

Dedicated to the memory of Egbertha Smit and Gerbrigje Boer

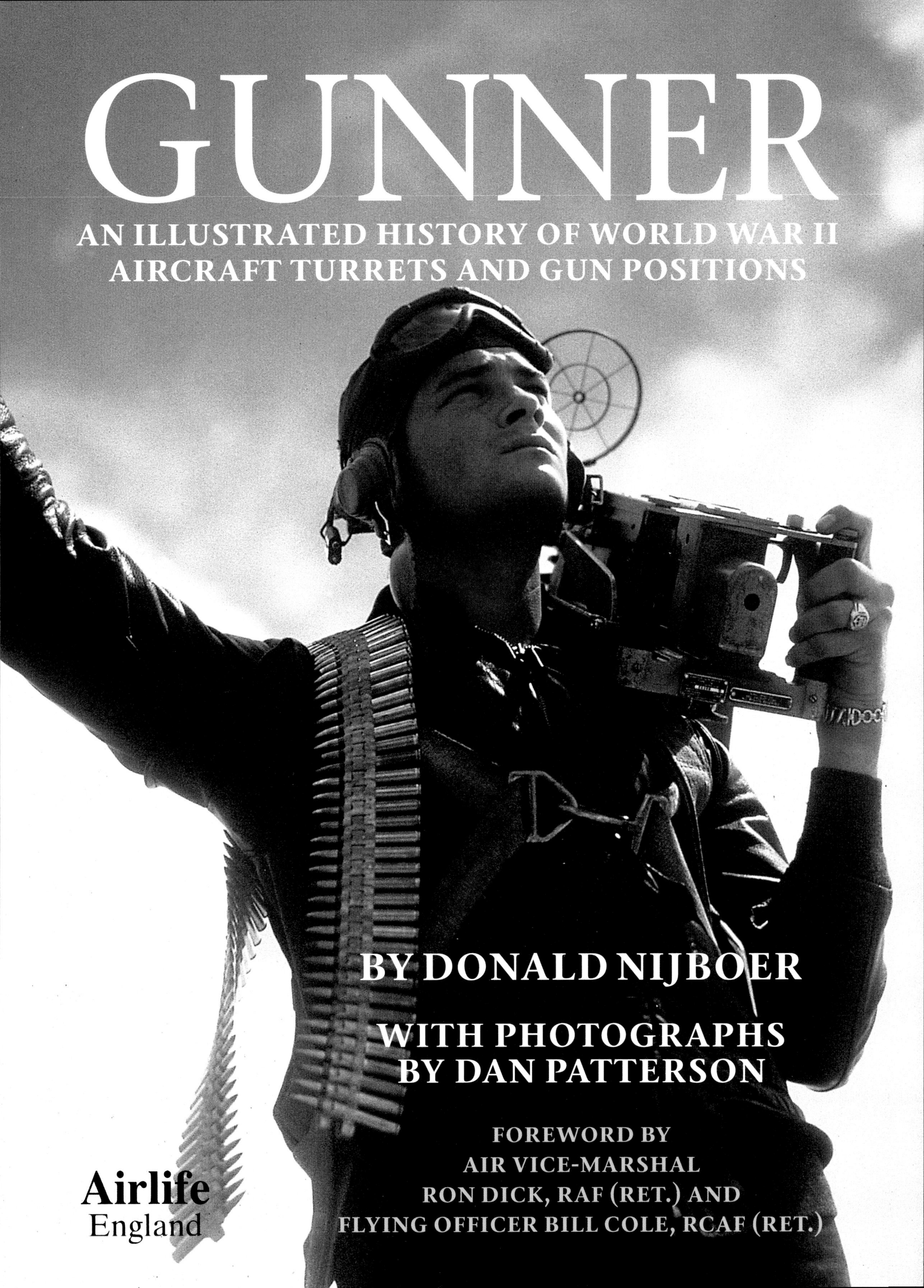

GUNNER

AN ILLUSTRATED HISTORY OF WORLD WAR II AIRCRAFT TURRETS AND GUN POSITIONS

BY DONALD NIJBOER

WITH PHOTOGRAPHS BY DAN PATTERSON

FOREWORD BY
AIR VICE-MARSHAL
RON DICK, RAF (RET.) AND
FLYING OFFICER BILL COLE, RCAF (RET.)

Airlife
England

First published in Canada in 2001
by Boston Mills Press
An affiliate of Stoddart Publishing Co. Limited

First published in the UK in 2001
by Airlife Publishing Limited

British Library Cataloging-in-Publication Data
A catalogue record for this book is available from the British Library
ISBN 1 84037 304 0

Design by Dan Patterson

Printed in Hong Kong

Airlife Publishing Ltd

101 Longden Road, Shrewsbury, SY3 9EB, England
E-mail: airlife@airlifebooks.com
Website: www.airlifebooks.com

HALF TITLE PAGE: The top of a B-17 Sperry ball turret along with items worn and used by a typical air gunner: Sheepskin B-3 jacket, A-4 gloves, B-2 flight cap, parachute, the powerful .50-caliber machine gun and the always present Varga pin-up girl.

TITLE PAGE: The romantic image of the air gunner was perpetuated throughout the war in advertising and publicity photos such as this one. The gunner's reality was a world of deadly fighters, flak, frostbite and death.

CONTENTS: Armed and ready, but with no enemy in sight. This trainee is ready for another practice flight aboard a North American O-47.

GUNNER

AN ILLUSTRATED HISTORY OF WORLD WAR II
AIRCRAFT TURRETS AND GUN POSITIONS

FOREWORD

Hate them or love them, respect or despise them, guns hold an awful fascination for people, to a degree unmatched by almost any other manufactured tool. Feared as instruments of death, they can also be admired for the craftsmanship that underlies their function. Their capacity to kill instantly and from a distance can give their users an almost godlike detachment from the act of killing, rendering it deceptively dispassionate and impersonal. This is particularly true for those whose occupations have guns at their core. Guns of one kind or another are necessarily central to the mission of all arms of the military, giving the armed forces the ability both to strike offensively and to defend. For the air forces of the 20th century, guns were, from the earliest days, essential to the conduct of operations. For fighter pilots, they became an extension of both body and will — the eyes focusing on the enemy, the attack conceived, the whole aircraft aimed as if it were itself a gun, the finger tightening on the trigger, the bullets carrying the mind's message of destruction to its desired conclusion. In some primeval sense, the victorious pilot gained in stature from the fight. A scalp was lifted and the pilot was honored, his name remembered.

Curiously, the same degree of celebrity was never attached to the gunners who defended bomber aircraft. Journalists and historians have little difficulty in naming the pilots involved in the significant bombing operations of World War II. Leonard Cheshire, Micky Martin, Curt LeMay, Paul Tibbets, Hansgeorg Bätcher — their names are prominent in the operational records of their services and in the historical accounts of their various feats. The same cannot be said of their crews. In recalling the story of the raid on the Ruhr dams, the pilots' names come quickly to mind — Gibson, Shannon, Hopgood, Martin, McCarthy, Maltby, and the others — but who remembers the name of Guy Gibson's tail gunner? Who was the man who fired the guns in Joe McCarthy's nose turret? Gunners have become perhaps the most anonymous warriors of World War II's bombing offensives. By far the most numerous members of aircrew and bearing the highest casualty rate, they are also the most forgotten. That is a pity, because in many ways they had the most challenging job of all.

There was nothing glamorous about manning guns in a World War II bomber, even though that was the period when air gunners had reached the apogee of their brief military history. As one RAF gunner has said, "The category of air gunner in the Royal Air Force lasted for less than fifty years. Post World War II, they are as outdated as the archers of Crecy."

Being a gunner was generally the sort of unrewarding task that had to be endured rather than relished. Frequently the youngest and lowest-ranking members of a crew, the gunners often found that their positions in the aircraft were remote from everyone else, and that they had to suffer long periods when the primary inputs to their senses were noise, cold and the constant expectation of being shot at. The quarters were cramped and the clothing was bulky. Loneliness and boredom combined to dull the awareness of men who were often little more than schoolboys. On many missions, they remained cocooned for hours in their private worlds, comparatively uninvolved in the business of getting the bomber to its target and delivering its weapons. In the face of flak, while the minds of most of the crew were kept occupied with the tasks of aircraft handling and navigating, the gunners could do little more than watch and wait, and feel helplessly inactive. Yet by day or night, they had to fight fatigue and stay alert, searching the skies for signs of the enemy. The nervousness they usually felt sometimes mounted into fear, but gunners were expected to react calmly and correctly under threat, instructing the pilot as to the whereabouts of fighters and advising on evasive maneuvers all while bringing guns to bear and making bursts of fire count. At such intense times, the other members of a bomber's crew knew that their lives depended on the good judgment of the man behind the guns.

Gunners took on their role facing appalling conditions, isolation, extreme hazard, and a loss rate high enough to discourage hopes of survival. That so many young men stuck to their guns and rose to the challenges of their dangerous job is a tribute to their spirit. Their courage was extraordinary. They deserve to have their names remembered.

AIR VICE-MARSHAL RON DICK,
CB, FRAeS, RAF (Ret.)

6

The motto "To serve and protect," used in policing today, could have served very well as the motto of the aerial gunners of both world wars. If you were an air gunner, your job was strictly to protect the aircraft and crew from the enemy and not to act as a fighter.

If you saw the enemy first, and he did not observe you, it was best to leave him alone and go about your business. Most gunners in Bomber Command never fired their guns in anger during night operations, but instead used evasive action such as the corkscrew to avoid an enemy fighter as soon as it was sighted before he saw you.

Author Donald Nijboer has outlined in detail the armament that were used on the many different aircraft types flown during the two world wars, particularly World War II, where more thought was given to the protection of the heavy bomber. It was during this time that the powered multi-gun turret came into its own. As World War II progressed, new, advanced turrets with heavier armament were introduced. By mid-1944, turrets with gun-laying radar known as "Village Inn," and twin .50-caliber machine guns were installed in the rear turrets of some aircraft. The new Rose turret, armed with two .50-caliber-machine guns, was also introduced. Arriving late in the war, these two new designs saw limited service and no real assessment of their value was ever made. When we heard of these new turrets, we were eager to try them out, but the war ended before they were put into full service.

During daylight operations, gunners and crews were at the mercy of enemy fighters. Using their heavy cannon armament, German fighters were able to attack from beyond the range of our turret armament. The American 8th Air Force tried to counter German daylight fighter attacks by using tight formations and heavy armament. Even with this strategy, the American losses were still quite staggering. It was not until fighter escort was provided that their losses came down.

The author has also brought attention to the difficulty gunners suffered in attempting to escape from the different gunner positions. Rear gunners had a better chance of escape as compared to the mid-upper gunner and the rest of the crew. Other crew members — the navigator, bombardier and flight engineer — had to pass through the aircraft to find the exit doors and jump. All the rear gunner had to do was open his turret doors, reach for his chest chute, which was hanging in the aircraft, hook it on, rotate the turret ninety degrees, and roll out backwards. It may sound very simple, but every second counted could mean the difference between life and death. Later on, we received seat pack parachutes, and that improved things considerably.

The photographs in this book show in striking detail the many different turrets and gun positions used during the Second World War. Here you get to see where gunners plied their trade and, through their experiences, find out what it was like to fly and fight in a cold, cramped turret high above Germany and the Pacific.

It was a time when the gunners were usually the "kids of the crew." Most of us were in our teens when we joined, eighteen or nineteen years old. We were looking for excitement and determined to do our part to put an end to the war. Many of those who joined paid dearly, and those of us who remain will never forget them. I wish to offer, on behalf of former air gunners, a special commendation to author Donald Nijboer and photographer Dan Patterson for this excellent presentation.

"We few. We happy few. We band of brothers."
— *HENRY V*, ACT V, SCENE II

BILL COLE, F/O REAR GUNNER,
428 GHOST SQUADRON, 6 GROUP, RCAF (RET.)

SIMPLE TRAINING FOR A DEADLY GAME.
AN INSTRUCTOR GIVING FINAL INSTRUCTIONS TO AIR GUNNERS OF THE ROYAL CANADIAN AIR FORCE BEFORE THEY BEGIN LIVE FIRING PRACTICE.

Introduction

When I was a kid I wanted to be a waist gunner. After watching *12 O'Clock High* with Gregory Peck, and the TV series, I thought the waist gunner had the best position in the B-17. From a visual point of view, it was the most exposed and exciting position. Other gunners were hidden in their turrets, but the waist gunner was out in the open. You could see him clearly, and I watched fascinated as he swung that big .50-caliber around and blasted away at enemy fighters. The noise and recoil of the .50-caliber machine gun was incredible. Spent shell casings and belt links poured out of the gun and fell to the aircraft floor as German fighters zipped by. The waist gunners fought back, protecting their aircraft and their buddies. It was no surprise that the good guys always won. But that was Hollywood. The reality, of course, was much different.

Turret development in the 1930s showed great promise. New, modern bombers were built with powered turrets installed. The belief was that the bomber with the new turrets could defeat any fighter attack and reach its target. The powered turrets were marvels of technology and design. Fully enclosed, they offered the air gunner double the previous firepower, protection from the elements and increased fields of fire. It was a giant leap forward, but one that was never fully tested. The theory of the self-defending bomber was shattered in combat in early 1939. The turret was not the answer to swift, heavily armed fighters and never would be. The British turned to night bombing to protect their bombers, while the Americans clung to their belief in the self-defending bomber well into 1943.

It's been said that the pilot's job was to get the aircraft in the air, but it was the gunner's job to keep it there. While many books and movies have been produced showing the exploits of pilots and squadron commanders, very few explore the important role the air gunner played. Most, if not all, air gunners were enlisted men with few privileges, low pay and little or no chance of promotion. While every member of the bomber crew played a vital role, it was the air gunner's responsibility to defend the aircraft and keep everyone safe. It was an awesome and indispensable responsibility. In the night skies over Germany, the gunner's vigilance was paramount for survival. A lack of concentration for only a moment could mean certain death by a prowling night fighter. There was also the constant danger of flak, collisions over the target area, and being hit by someone else's falling bombs. Air gunners suffered through long periods of extreme cold, noise and constant shaking.

In daylight raids American gunners could see and fire at their attackers. Those gunners had to deal with persistent head-on attacks from German and Japanese fighters, and once on the bomb run they were unable to dodge the heavy, dirty flak. Exploding aircraft often sent jagged pieces of metal flying through their tight formations—along with personal items and human body parts. The air war was a gruesome one, and in many ways more grueling than the infantryman's.

Gunner picks up where *Cockpit: An Illustrated History of World War II Aircraft Interiors* left off. Through Dan Patterson's remarkable photography we get to see up close the cramped, cold, exposed turrets that so many young men fought and died in. Veterans describe what it was like to fly and fight in the many different turrets and gun positions. Illustrated diagrams show the inner workings of some of the more well known turrets. *Gunner* not only reveals, for the first time in print, the turrets of famous aircraft such as the Lancaster and B-17, but it also features some extremely rare gunners' positions from aircraft such as the Me 410, the Italian S.M. 79, and Russian Il-2 Shturmovik. It's a chance to see what many of us have never seen, and maybe for a brief moment to imagine what it must have been like to be an air gunner in the Second World War.

Donald Nijboer

"At the going down of the sun, and in the morning we will remember them."
— Lawrence Bynon

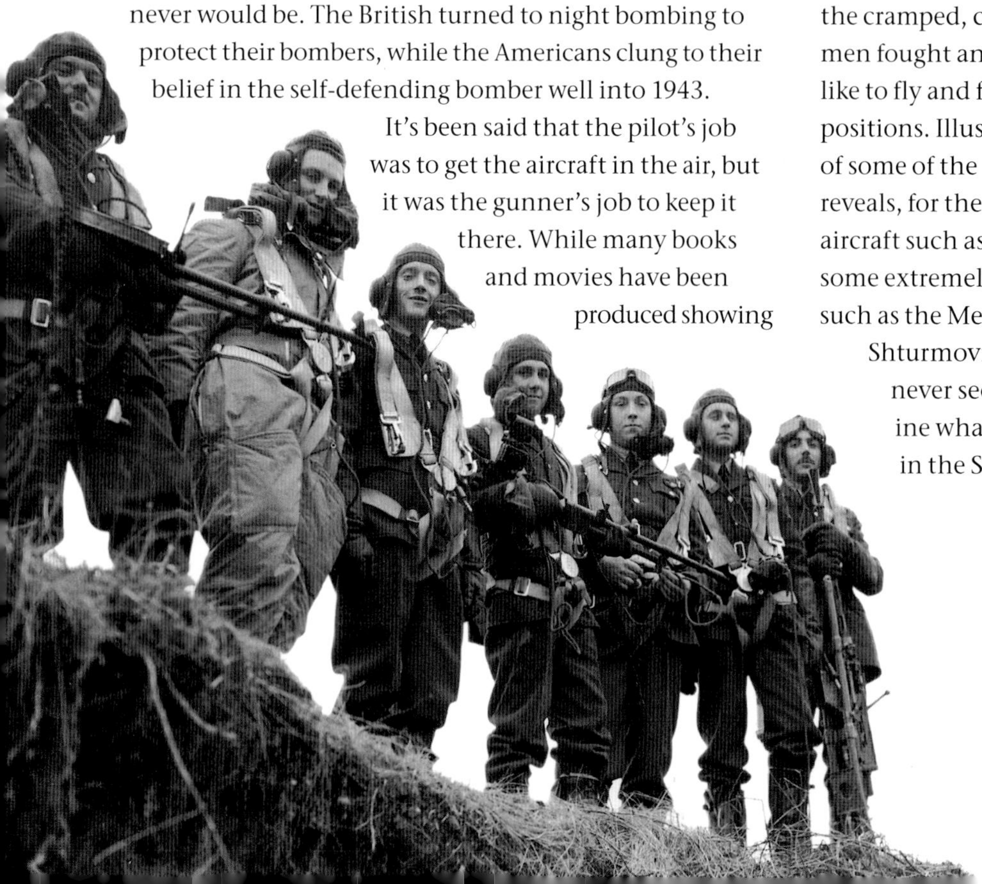

Before Donald Nijboer and I finished work on our previous book, *Cockpit: An Illustrated History of World War II Aircraft Interiors*, we realized that there was a large empty space on the bookshelves where the book you're holding belongs. The great majority of airmen from all the fighting nations of World War II flew and fought and died in areas of the warplane other than the cockpit. Their world was often a very dangerous one. These men spent hours encased in steel and Plexiglas bubbles or peering into blinding skies through open windows, frigid blasts of high-altitude air powered by the slipstream hitting them right in the face. The men in the cockpit at least had the mental advantage of being in control of the warplane; the gunners had to put their faith — and often, fate — in the hands of the pilots.

Aerial warfare in the 1940s generally called for large numbers of multi-engined aircraft to fly in formation at relatively slow speeds, over long distances, to attack by bombing a single target. The opposition would defend their homeland with swift and agile single-engined fighters armed with increasingly heavy weapons designed to destroy the bombers. There were, of course, exceptions and experiments, but this was mostly the rule. This was to be the combat proving of the strategic bombing theories that were written between the world wars. How effective the strategies were remains a topic of discussion now, sixty years later. What cannot be denied is the cost of aerial warfare in the Second World War. The United Kingdom's Royal Air Force Bomber Command alone suffered over 50,000 casualties; the American 8th Air Force had over 30,000 killed. The greatest number of those losses were gunners. In an American B-17 or B-24, the crew of ten men included two pilots, a navigator and bombardier; the rest were gunners. An RAF Lancaster carried a crew of seven — a single pilot, a navigator flight engineer, wireless operator, bomb aimer and two gunners.

This book is about the rest — the men who learned to shoot at moving targets from a moving platform, at high speed and altitude. These were soldiers forced to face terrible odds of survival in a highly dangerous combat environment, where there was no place to crawl into for protection when attacked. They had to be able to repair damaged guns and turrets, give first aid to their wounded crewmates, and they had to have the nerve to clip a very small parachute to a webbed harness and throw themselves out of a damaged or burning bomber, trusting that it would work properly. These were the gunners.

I have unending respect for these men.

DAN PATTERSON

F/SGT. GOLDSMITH'S STATEMENT OF THE ATTACK BY FIGHTERS
OVER BREST ON FORTRESS I OF 90 SQUADRON,
POLEBROOK

on the 16th August,1941 (M.S.I. MUD 8577 - Note:-
F/Sgt.Goldsmith was fire controller in
Fortress.

Sighted enemy fighters at 32,000 feet.

"Ack,"Ack" dropped away directly below.(very accurate and intense)

Two aircraft approaching Port Quarter 5,000 ft. below. These fighters were very fast on the climb.

Order guns to "Stand By".

Enemy aircraft reached Fortress' height and attacked in formation "Dead Astern" closed to 300 yards Fire Controller hit in first attack.

Identified fighter as He.113a, camouflage - all black excepting for tail fin which was painted yellow.

Four attacks were made from "Astern", average closing range 300 yards, all attacks made in formation. Vapour trails were continuous throughout both fighters and bomber.

Sgt. Leahy hit in the third attack,Sgt.Ambrose unable to fire owing to vapour trails (beam gunner). Sgt.Leahy did not say that he was hit, but his microphone was switched on and the "Fire Controller" could hear him groaning.

During the combat the Ack Ack ceased completely.

The enemy aircraft attacked again, broke away beneath and climbed up to attack from Astern and slightly below.

Evasive action was carried out from Port to Starboard by turning, height was maintained throughout the engagement.

The Fortress was hit several times in this attack with cannon and machine gun. Combat lasted 7 minutes.

He.113s broke away and two 109Fs attacked immediately from Starboard Quarter.

Sgt.Ambrose was hit seven or eight times in this attack. Sgt.Needle was unable to get his guns to work(freezing suspected). Sgt.Needle immediately on request from the Fire Controller sent out an S.O.S. which was received in this country.

109Fs closed to 150 yards firing from Starboard and Port quarters, Sgt.Needle again manned the mid-upper guns but was hit by a cannon shell. The 109Fs broke away and two more 109Fs attacked, height still being maintained 32,000 feet. Evasive action taken by turning. Hits were obtained in the wing and fuselage. The 109Fs then delivered attacks from above on the wide Quarters, Astral Dome hit by a cannon shell.

Attack broke off at 32,000 feet, Duration of combat, 12 minutes, Camouflage of 109Fs - all black, excepting nose and tail fin which was yellow.

Last attack delivered immediately the previous two 109Fs broke off by one 109F. In this attack Fortress made for sea level in a straight dive.

THIS CHILLING ACCOUNT BY FIRE CONTROLLER F/SGT. GOLDSMITH REVEALS
HOW VULNERABLE BOMBER CREWS WERE WHEN FACED WITH DETERMINED FIGHTER ATTACKS.

The camouflage of this 109F was as follows:- Yellow nose and tail fin. Black fuselage with a red crest on each side.

Me.109F attacking from Quarter Astern to 8,000 ft. closing to 100 yards. Hits were obtained on Fortress. Attack broke off and Fortress levelled out at approximately 2,500 ft.

Total time of combat 23 minutes.

Total number of attacks 26.

GENERAL.

After the combats broke off the Fire Controller looked aft and noticed Sgt.Needle in a kneeling position with his hands clasped to his stomach.

Sgt.Leahy was still groaning and Fire Controller attempted to render first aid, but was unable to do so, as the bomb compartment was opened and the rope rail was missing. The Fire Controllers leg was useless.

Observer tried to get through, but was unable to do so. During this period and previous period petrol was pouring out of the Starboard inner tanks.

Vapour trails started at 25,000 feet and continued until Fortress dived for the sea level.

Enemy aircraft had vapour trails. Engines were running well throughout engagements, until reaching English coast when aircraft became unstable, rudder was smashed and wings and fuselage peppered with cannon and machine gun.

The He.113s seemed to have a greater rate of climbing than the 109Fs.

109Fs favoured attacking from Starboard Quarter, this may have been due to the fact that in taking the correct avoiding action the Fortress would be heading back towards the enemy coast. The Fire Controller counteracted this by turning and straightening out immediately, thus heading in the general direction of the English coast.

Astral Dome frozen.   Oxygen O.K.

Temperatures not too extreme.

Note:- This statement obtained by F/L. Meyer, Gunnery Leader of

90 Squadron. 21.August,1941.

A DRAMATIC SHOT OF A B-17E RUNNING UP ITS ENGINES AT TWILIGHT. THE B-17E WAS THE FIRST FULLY COMBAT-READY VERSION OF THE FORTRESS TO ENTER SERVICE WITH THE USAAF.

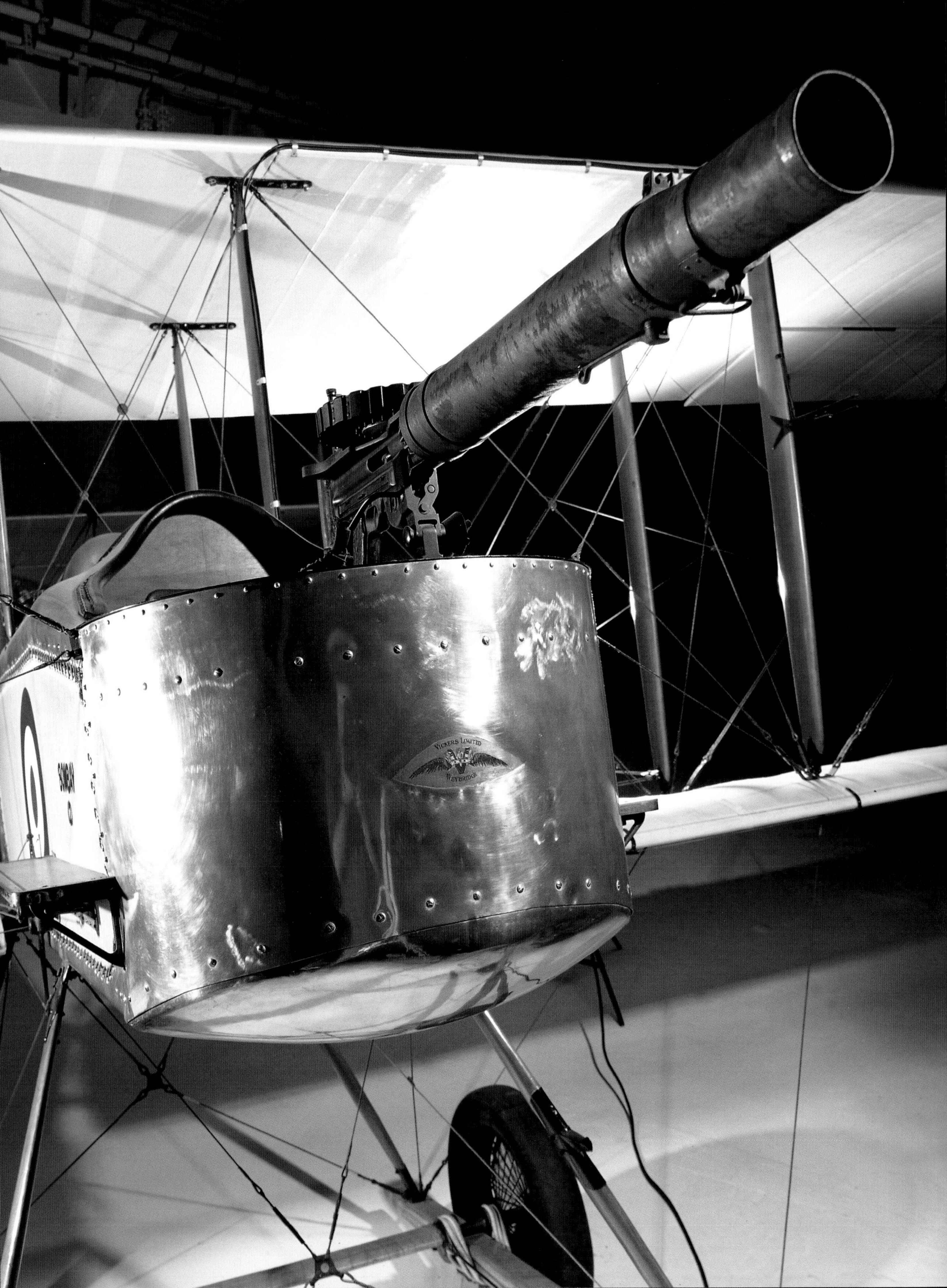
BOMBAY
VICKERS LIMITED
WEYBRIDGE

BACKWARDS INTO BATTLE

Little more than eight years after the Wright Brothers freed man from gravity's grip, the air gunner was born. On June 7, 1910, Lieutenant Thomas de Witt Milling and Captain Charles De F. Chandler proved that man's inventiveness was as strong as his desire for destruction. As the two sat side by side in the Model B Wright Flyer, Chandler made the final adjustments to the Lewis gun strapped between his legs. After a successful takeoff, Milling leveled the aircraft and flew toward the canvas target. Chandler gripped the Lewis gun, took aim and pulled the trigger. At that moment everything changed. The aircraft armed with the machine gun would rule the high ground and would soon become one of the most destructive instruments of war ever invented. This was the first experiment in which a large, heavy-caliber machine gun was carried aloft. The results were impressive. At an altitude of just 300 feet, Chandler fired a full drum of forty-seven bullets at a canvas target below. Roughly twelve percent of those bullets found their mark.

European tests soon followed, and on November 27, 1913, pilot Marcus D. Manton and gunner Lieutenant Stellingwerf of Belgium fired a Lewis gun from the air. Again, hits were registered. From the first burst of twenty-five rounds, eleven found their mark. From his second burst of forty-seven rounds, only fifteen bullets registered. The Lewis gun, the standard infantry machine gun of the time, weighed 26½ pounds and had a rate of fire of 560 rounds per minute. With minor modifications, this weapon would serve through the First World War and continue well into the Second World War.

Early airplanes were extremely fragile machines. Most could not carry a pilot to any worthwhile height, let alone a machine gun as well. At the outbreak of the First World War, most countries equipped with aircraft had very few that were designed to carry guns. One exception was the British Vickers FB5 Gunbus. The original Admiralty contract called for a "fighting aeroplane" armed with a machine gun. The addition of a machine gun called for a "pusher"-type aircraft with the crew positioned forward

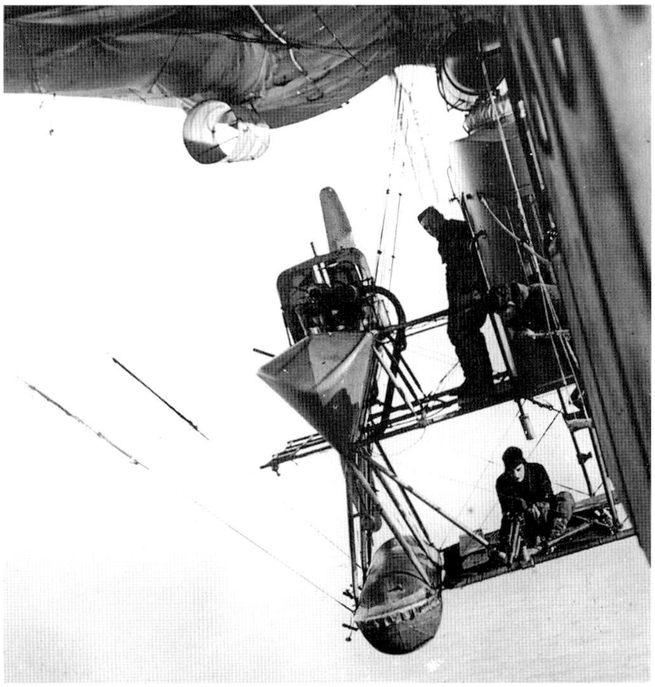

of the engine. Until the introduction of synchronization gun gear, the pusher-type aircraft was the most effective way to mount a machine gun. As the war progressed, aircrews made strenuous efforts to fit more machine guns to their already overburdened machines. Their maximum speed rarely climbed above 70 mph and the service ceiling bumped no higher than 9,000 feet!

Seated in the front cockpit, with the pilot located right behind, the observer/gunner occupied an extremely precarious position. The cockpit was extremely small and made of plywood. Despite the fact that they were exposed to wind, rain and the ever-present cold, gunners flying in the FB5 still managed to claim a number of successes. When you consider that these men were held in by just a single strap and were constantly exposed to the elements, it is a wonder they hit anything at all.

The first air-to-air kill using a machine gun was credited to an observer/gunner in 1914. Using a Voisin pusher-type aircraft, Corporal Louis Quenaults, with pilot Sergeant Joseph Frantz at the controls, managed to shoot down a German Aviatik two-seater. As the war progressed, German and French aircraft began to appear with "free"

LEFT: THE VICKERS FB5 GUNBUS WAS ONE OF THE FIRST TRUE FIGHTING "AEROPLANES." FITTED WITH THE STANDARD LEWIS INFANTRY MACHINE GUN, ITS FRONT POSITION OFFERED LITTLE IN THE WAY OF PROTECTION FOR THE OBSERVER/GUNNER.
RAF MUSEUM, HENDON.

ABOVE: THE AIR GUNNER SERVED IN MANY DIFFERENT PRECARIOUS SITUATIONS. THIS PHOTOGRAPH OF A ROYAL NAVY AIRSHIP GUNNER CLEARLY SHOWS WHAT AN EXPOSED AND VULNERABLE ROLE HE PLAYED.

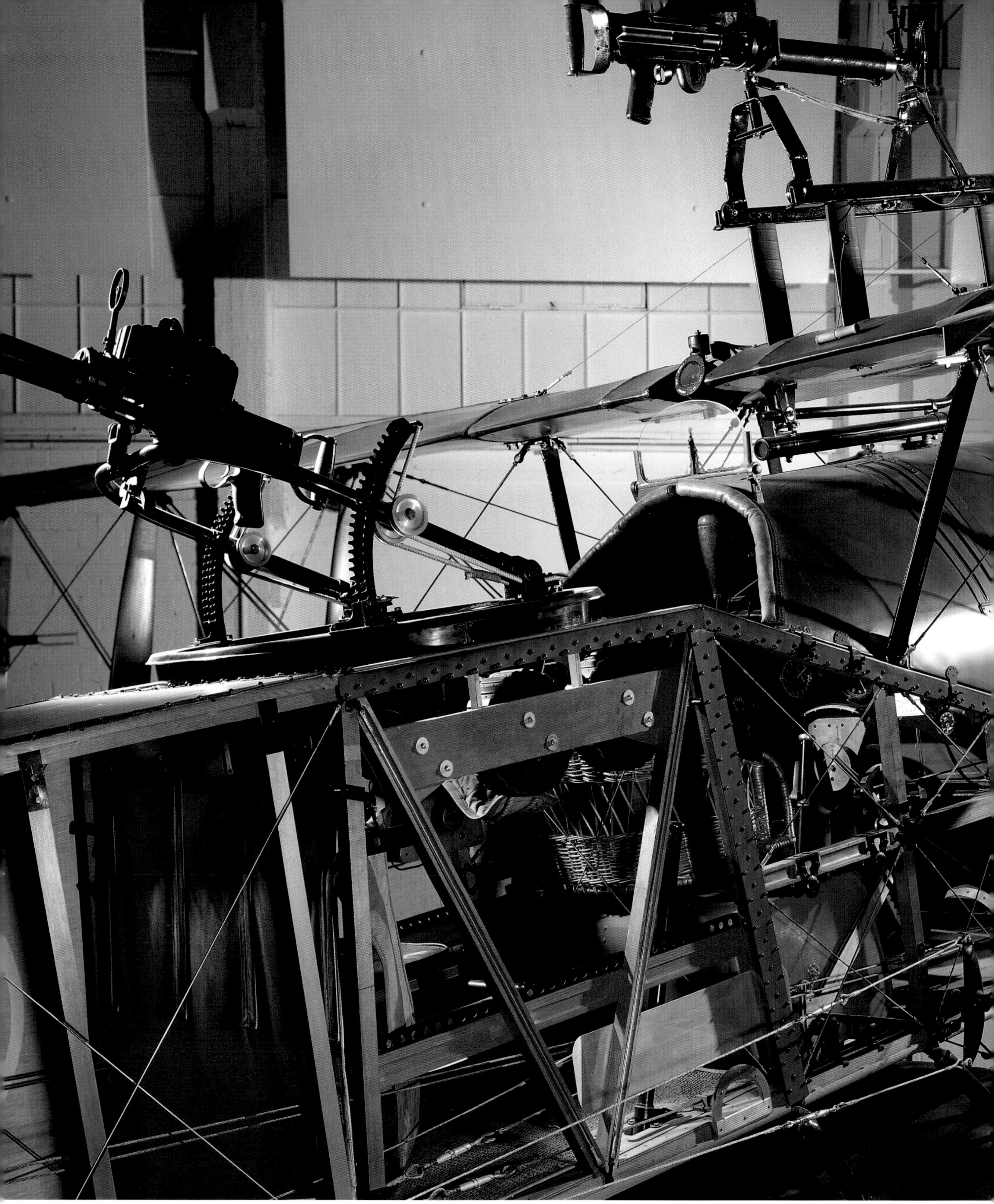

The Bristol F2b was one of the most heavily armed two-seater fighters to fly in World War I. It was armed with a Mk II Lewis gun in the rear gunner's position, a Mk III mounted on the upper wing and a fixed Vickers that fired through the propeller. The Scarff No. 2 ring, clearly shown here, became the standard observer/gunner's mounting throughout 1916–18.
RAF Museum, Hendon.

gun mountings on two- and three-seat reconnaissance aircraft. These were primarily for defense, but on occasion were used successfully in air-to-air combat.

As the role of the airplane quickly evolved, the responsibilities of the gunner remained unchanged. New gunsights, specific training and new ammunition were hardly considered. The real problem of hitting a fast-moving airplane from another equally swift machine was in many ways ignored. Gunners learned on the job. At the time, British "back-seaters" were mostly infantry officers who volunteered for transfer to the air arm. Finally, in August 1915, the Royal Flying Corp realized the need for proper training and laid down the qualifications needed to become an observer/gunner. Unlike their World War II brethren, World War I gunners had to fill many roles. Not only did they have to defend their aircraft, but they also had to have a proficient knowledge in the use of RFC cameras, be able to send and receive wireless messages at a rate of six words a minute, and possess an excellent knowledge of airplane-artillery co-operation methods and a thorough knowledge of the Lewis gun.

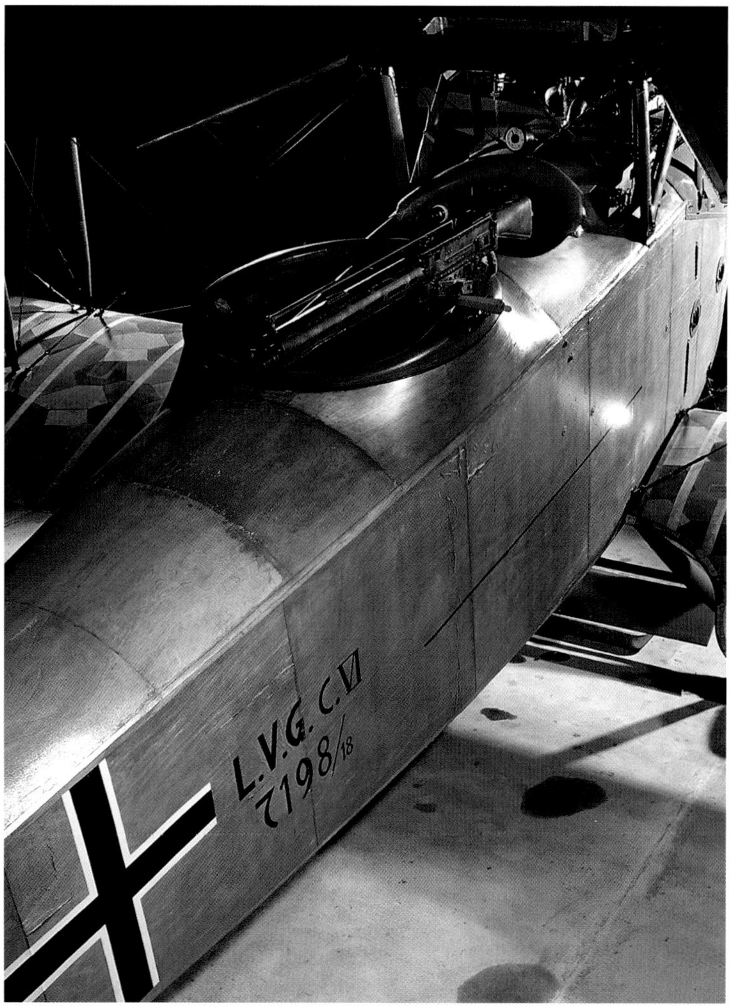
L.V.G. C.VI
7198/18

By 1917 the pusher-type aircraft had been surpassed by the tractor type, with the engine in the front. One of the best two-seater tractors of the war was the Bristol F2b fighter. Nicknamed "the king of the two-seaters," the F2b quickly established itself as one of the most successful fighters of the war. The combination of pilot and gunner gave the F2b a unique and extremely effective one-two punch. Other multi-seat aircraft to enter service at the time included the de Havilland DH 4 day bomber, the Handley Page 0/100 bomber, and the giant twin-engine German Gotha bomber. The Handley Page 0/100 and Gotha were similar in size and usually carried three gunners. Even with multiple gunners, the bombers suffered when faced with determined fighter opposition. The British DH 9, introduced in 1918, was never popular and was totally outclassed by German fighters. In five months of service with the Independent Forces, DH 9 observer/gunners managed to shoot down 157 enemy aircraft. They lost 246 observer/gunners through fatalities, wounds or injuries. Victory tallies were not credited to observer/gunners. Little publicity was given to the back-seaters, even though a number of aces, including Major W. A. Bishop, Major W. G. Barker and Major J. T. B. McCudden, were initiated into aerial warfare via the back seat. Many of the gunners' victories were credited to or "shared" with the pilot.

In March 1917 the Germans began to bomb Britain with their large twin-engine Gothas. Using auxiliary fuel tanks and oxygen for their crews, they reached London and caused a surprising amount of damage. Britain's fighter defense at the time consisted of old second-line aircraft, which were unable to hurt the bombers. Later model fighters were rushed home from the front to deal with the new threat. The London Air Defence Area was quickly established, and the new defenses shot down enough bombers to force the Germans to switch to night bombing.

At the end of the war, the observer/gunners' equipment had changed little. For the most part, the multi-seat gunner's cockpit was still open to the elements with no protection against the icy slipstream and paralyzing

THE GERMAN 7.92-MM PARABELLUM LMG 14 MACHINE GUN WAS AN EXCELLENT WEAPON. BELT-FED FROM A 250-ROUND DRUM, THE LMG 14 OFFERED MORE SUSTAINED FIREPOWER THAN THE 47- AND 97-ROUND MAGAZINES CARRIED BY THE BRITISH LEWIS GUN. THE SHUTTLEWORTH COLLECTION.

Labels within image:
1st Position. under fire
2nd Position. diving down still exposed to fire
4th Position. Gun out of action
5th Position turning & diving to bring Gun into action again
1st Position
2nd Position
3rd Position
3rd Position. diving under opponent to dodge his fire
PILOT
GUNNER & OBSERVER
Machine gun swivelled up & down to give fan like spread to bullets

THIS *LONDON ILLUSTRATED NEWS* DRAWING CLEARLY SHOWS HOW CRUDE EARLY AIR-TO-AIR FIGHTING REALLY WAS IN WORLD WAR 1. RUDIMENTARY GUNSIGHTS DIDN'T ALLOW FOR ACCURATE SHOOTING. THE BEST CHANCE AT SUCCESS WAS TO FILL A PATCH OF SKY WITH AS MUCH LEAD AS POSSIBLE AND HOPE FOR A HIT.

cold at altitude. The one thing that was standardized was the use of the Scarff No. 2 ring mounting. It would remain in use in the Royal Air Force for more than twenty years. Parachutes were rarely worn and, in most cases, nonexistent; oxygen use was only in the experimental stage; and protective clothing consisted of layers of leather, fur and canvas. The basic tools of the trade remained unchanged. The rifle-caliber machine gun that entered service in 1914 was still in use. Twenty years later, the gunners of the Second World War would find themselves facing some of the same problems, but help was on the way.

The impact of the airplane on the outcome of the First World War was minimal. Trench warfare, in all its ugly forms, continued for four straight years without change. But the flying machine was evolving. In four short years, aircraft went from flimsy machines looking for a part to play to efficient, specialized killing machines. At the end of hostilities, each side was equipped with aircraft that were designed to fill a specific role: fighter, fighter bomber, ground attack, torpedo bomber, anti-submarine, reconnaissance, and strategic bomber. The end of the war also saw the rapid reduction of the world's air forces. In just eighteen months, the Royal Air Force went from one of the world's largest to a mere shadow of itself. For the next fifteen years the RAF would soldier on with First World War aircraft such as the de Havilland DH 9 and the Bristol F2b fighter.

During that time, RAF aircrew fell into two categories: pilots or observers. The observer's role had changed little from the First World War, and in the RAF, observer/gunners were mostly volunteer ground tradesman who were eager to fly.

Funding was another major problem, and with most air forces fighting the other services for cash, it's no wonder most were small and ill-equipped. But with new technology came new ideas. The German bombing of London during the First World War had introduced a new kind of warfare — strategic bombing. Men such as U.S. Brigadier-General Mitchell and

TOP: THE REAR GUNNER'S POSITION IN THE CAPRONI CA 36 BOMBER WAS THE MOST EXPOSED OF ANY BOMBER THAT SERVED IN THE FIRST WORLD WAR. IT WAS ARMED WITH THREE FIAT REVELLI 6.6-MM MACHINE GUNS. USAF MUSEUM, DAYTON.

BOTTOM: THE ONLY EXISTING TWIN-ENGINE GERMAN BOMBER TO SURVIVE THE FIRST WORLD WAR IS AT THE NATIONAL AVIATION MUSEUM, OTTAWA. THIS A.E.G III IS ARMED WITH TWO PARABELLUM MACHINE GUNS.

Britain's Chief of Air Staff, Major General Sir Hugh Trenchard, believed that the best defense against an enemy was to attack from the air. The future lay in building a heavy bomber force that could strike at the heart of the enemy and destroy its capacity to wage war. The ideas were new and bold, but the aircraft designed and built at the time could not do the job. Nor were there enough of them. The Handley Page Heyford, which first flew in June 1930, was at the time the RAF's most advanced bomber. The last biplane heavy bomber to enter service, it saw frontline service until 1937!

"The mid-upper gunner in the Heyford was protected from the slipstream by a hinged board about nine inches high. But as soon as you put it up, the airspeed dropped by five miles an hour and the pilot noticed immediately! Of course, they didn't like you standing up in the gunner's cockpit, but you had to because that was the only way you could fire!"

SERGEANT JOHN HOLMES,
AIR OBSERVER

TOP: THE MARTIN B-10 INTRODUCED THE FIRST FULLY ENCLOSED TURRET. IT WAS NOT POWERED AND WAS FAR FROM PERFECT. USAF MUSEUM, DAYTON.

BOTTOM: THE DORSAL GUNNER OF THE B-10 STILL BORE A STRONG RESEMBLANCE TO HIS WORLD WAR I BRETHREN. HIS JOB WAS MADE EVEN MORE DIFFICULT WITH THE INCREASE IN SPEED AND ALTITUDE OF THE NEW MONOPLANE BOMBER.

In the United States a revolution in bomber design was about to take place. In 1932 the Glenn L. Martin Aircraft Company introduced the Martin B-10 monoplane bomber. The B-10 was a quantum leap forward in design and technology. It was the most modern bomber in the world and would establish the pattern for everything that followed. The B-10 was an all-metal monoplane with retractable undercarriage, flaps, variable-pitch airscrews and the first fully enclosed turret. The B-10 carried a crew of four and was armed with three .30-caliber machine guns: one in the nose, and one each in upper and floor positions. The nose turret was manually operated. Headroom was inadequate, and the open gun slot caused the slipstream to roar through, forcing the gunner to wear his goggles at all times! It was not a perfect solution, but it was described as easy to operate at high speeds. With a maximum speed of 213 mph, the B-10 was 60 mph faster than the Heyford and could carry a 2,000-pound bomb load to a maximum range of over 900 miles. U.S. fighters were still of the biplane variety and could not intercept the new bomber.

The introduction of the B-10, along with the development of the Norden bombsight in 1932, affected both U.S. fighter and bomber aviation. As each new bomber design was introduced, with increased speed, range and general performance, fighter aircraft suffered. They lacked parallel developments and soon became less important. This helped foster the idea that the bomber did not need fighter escort and that any large bomber formation would be able to defend itself. Brigadier-General Oscar Westover, Assistant Chief of the Air Corps, stated that "bombardment aviation has defensive firepower of such quantity and effectiveness as to warrant the belief [that] with its modern speeds it may be capable of effectively accomplishing its assigned mission without support." (This belief would hold well into 1943, when American bombers would strike deep into Germany with serious losses.)

As aircraft speeds increased, so did the need to protect the gunners. Most if not all the defensive guns used in World War I were open to the elements, with no protection from the slipstream. It took great physical strength to aim the gun and score hits. Gale-force winds and extreme cold slowed the gunner's reflexes. Lack of oxygen at high altitude, vertigo and nausea were common. Add to that the weight of the heavy clothing required, and you can imagine how ineffective a gunner might be if attacked by enemy fighters.

THE BOULTON PAUL OVERSTRAND WAS THE FIRST BRITISH BOMBER TO BE EQUIPPED WITH A POWERED TURRET. ARMED WITH A SINGLE LEWIS GUN, THE BOULTON PAUL TURRET WAS CAPABLE OF ROTATING A FULL 360 DEGREES WHEN THE GUN WAS ELEVATED TO 70 DEGREES. BOULTON PAUL GUNNERS ACHIEVED A HIT RATE OF 55 PERCENT AGAINST TOWED TARGETS, COMPARED TO 15 PERCENT BY GUNNERS USING SCARFF RINGS IN OPEN COCKPITS.

New bombers such as the B-10 and Boulton Paul Overstrand began a trend that would see the development and widespread use of the powered turret in World War II. In Britain, companies such as Boulton Paul, Nash and Thompson, and Bristol began design work on the enclosed, powered turret. The first powered turret to see RAF service was a Boulton Paul design. In 1935, 101 Squadron was the first to be equipped with twin-engine Overstrands armed with nose-mounted powered turrets. The turret was a great success, but it was not perfect. Due to its slow rate of rotation and the compressed air required to move it, the turret was limited in its application. This problem was solved by the adoption of a system developed in France. French engineer J. B. A. de Boysson, of the Societe d'Applications des Machines Motrices, had built a turret that combined electrics and hydraulics. The self-contained hydraulic generator was powered by an electric motor that was confined to the rotating portion of the turret. This was a great breakthrough. This self-contained unit provided the gunner with a comfortable seat and gun apertures that were sealed, preventing the cold slipstream from blasting through. Remarkably, the French government and aircraft industry were not interested in the design. The British were more than happy to acquire the license for the turret's development, and quickly instructed other firms to produce competing designs. Soon, powered turrets from Bristol, Nash and Thompson, and Boulton Paul were standard equipment on British bombers.

In 1934 Italian aeronautical engineer Gianni Caproni developed and built upper and lower turrets for the Caproni Ca 89. These turrets were manually operated, but by 1937 powered turrets were in use on the majority of Italian bombers. Surprisingly, Germany's rearmament program did not include powered turrets. Even with major advances in airframe and power-plant development, German bomber armament centered on the hand-held machine gun. It was believed that speed and hand-held guns could defeat any fighter attack. In the United States, turret development before 1939 was slow to nonexistent. Even the famous Boeing B-17 was not equipped with turrets. The prevailing attitude at the time was that turrets were "too heavy and complicated."

When war broke out in 1939, the Royal Air Force had the only strategic bomber force equipped with the most advanced powered turrets in the world. As impressive as that might sound, only five squadrons of Vickers Wellingtons (Frazer Nash nose and tail turrets) and nine squadrons of Armstrong Whitworth Whitleys (Frazer Nash nose and tail turrets) were available for combat. The rest of Bomber Command was equipped with fifteen squadrons of the obsolete Fairey Battle, eight with the Bristol Blenheim, and six with the Handley Page Hampden.

THE ORIGINAL, OFFICIAL CAPTION FOR THIS PHOTOGRAPH READ: "HITLER WOULD LIKE THIS MAN TO GO HOME AND FORGET ABOUT THE WAR. A GOOD AMERICAN NON-COM AT THE SIDE MACHINE GUN OF A HUGE YB-17 BOMBER IS A MAN WHO KNOWS HIS BUSINESS AND WORKS HARD AT IT."

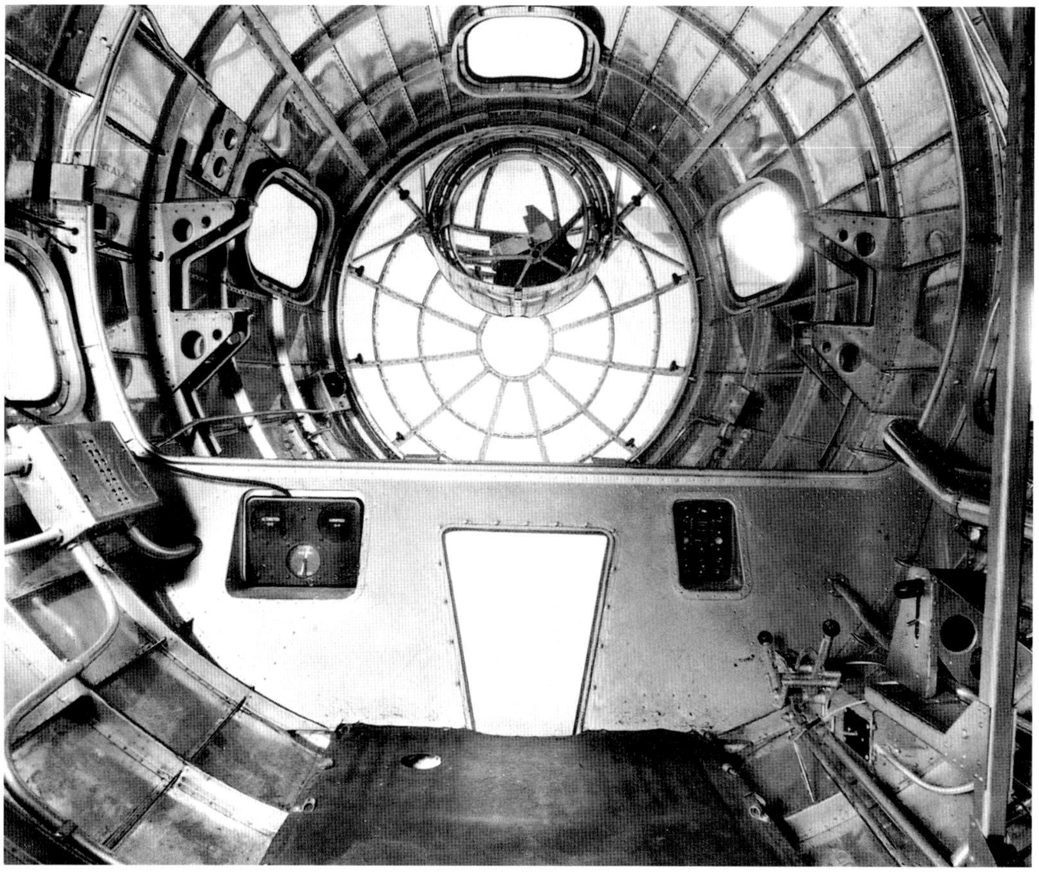

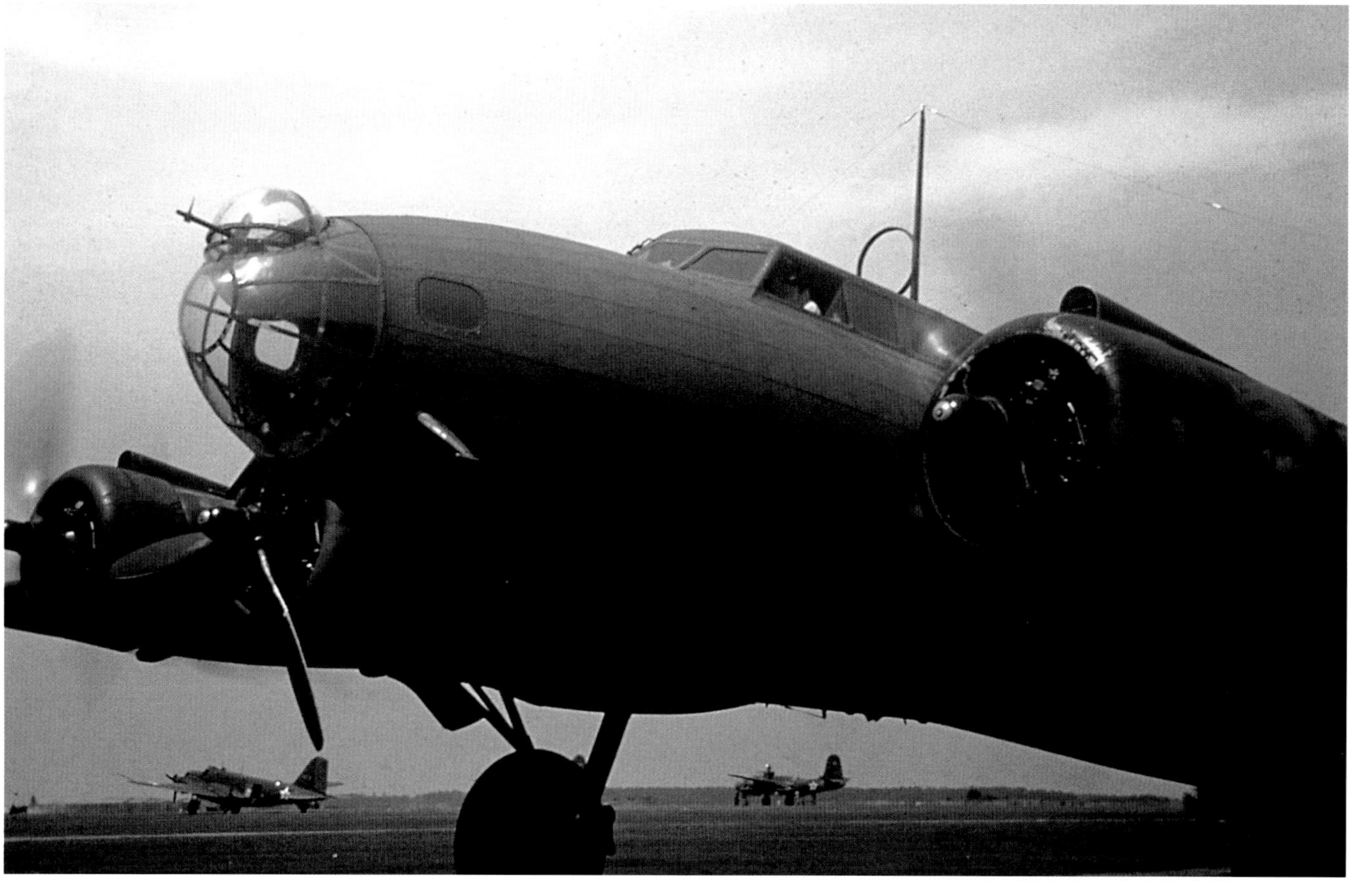

Top: The nose gunner's position in the YB-17.
Bottom: Although christened the "Flying Fortress" by the press, the YB-17 was lightly armed and carried no powered turrets.

RUNNING THE GAUNTLET

The advent of the powered turret finally elevated the role of the air gunner to a new status. In the era of the open cockpit, he was regarded simply as a passenger, ballast to balance the aircraft's center of gravity. The new, sophisticated powered turret with double the firepower and increased accuracy was now an integral part of the bomber operations. In 1939 the British Air Ministry recognized the air gunner as aircrew in two categories: air gunner (AG) and wireless operator / air gunner (WOPAG), later shortened to wireless air gunner (WAG).

For the air gunner entering World War II, the future was not promising. Even with the introduction of the powered turret, the air gunner faced a world of cold, numbing isolation. Although he was fully equipped with a complete kit of flying clothing, there was no provision for heating in the turret. Nor were his guns and ammunition supply protected from the severe arctic conditions at high altitude. Frostbite was a constant enemy, but not the only problem. High-altitude flight meant a secure supply of oxygen was vital. Without it, anoxia resulted and, if not caught quickly, led to euphoria and death. Many gunners perished due to damage to or failure of the oxygen supply. Without oxygen, the gunner's remaining lifespan was measured in minutes. Working in a cramped, unheated turret at 20,000 feet, for five, six, eight, ten hours or more in a high state of readiness required extraordinary physical endurance, not to mention mental toughness. But the greatest hazard (apart from fighter attack and flak) was

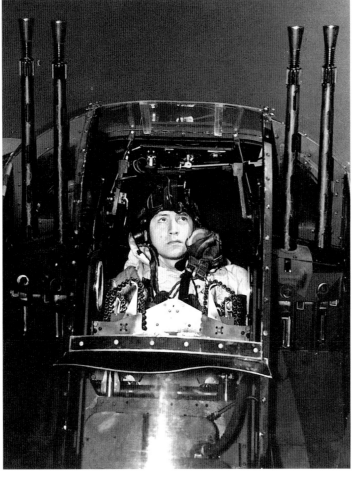

fire. Surrounded as he was by hydraulic pipelines, electrical wiring, oxygen lines, and explosive ammunition, the air gunner's chances of survival in a burning aircraft were minimal. The very design of the aircraft fuselage caused any fire in the aircraft to blow backward. This slipstream-driven furnace burned and killed many gunners. The isolated top, waist and rear turret gunners were on their own. It was not a question of just jumping out. For example, the tail turret gunner's parachute was stowed outside of the turret behind the turret doors. In order to bail out, the gunner had to re-enter the burning aircraft, strap on his parachute, find the escape hatch and jump. Heat damage also caused turret doors to jam shut, cutting the gunner off from any means of escape.

When war broke out in 1939, most if not all the bombers, on both the Allied and Axis side, used turrets and free-mounted machine guns as defense. The de Havilland Mosquito, which had no defensive armament, was the exception — and ironically, it finished the war with loss rate of just 0.65 percent per sortie (compared to 2.20 for the Lancaster). The RAF held to the belief that with the new powered turret their bombers could fight off any fighter attack, strike the target, and return to base. Early daylight RAF attacks on German soil were directed at naval targets. The results were not promising. Raids undertaken by Blenheims and Hampdens (the Hampden had no powered turrets) suffered heavy losses, but this was attributed

THE FOUR-GUN TURRET WAS STANDARD ON BRITAIN'S FOUR-ENGINE HEAVY BOMBERS.
MANY GUNNERS CUT OUT THE FRONT PERSPEX FOR BETTER VISIBILITY.

to bad formation-keeping and inadequate defensive fire-power. On December 3, 1939, twenty-four Wellingtons attacked German warships. Bf 110s and Bf 109s attacked the Wellington formation but were unsuccessful, but one 109 was hit and possibly shot down. The raid seemed to confirm the belief that turret-equipped bombers would always get through. The euphoria was short lived.

On December 18, a force of twenty-four Wellingtons set out to bomb German warships off Whilhelmshaven. The weather was excellent and visibility was perfect. An experimental Freya radar station detected the Wellingtons at a distance of 70 miles. German fighters were quickly scrambled and the battle began. Twelve of the twenty-two Wellingtons that reached the target area were shot down. Six more crash-landed after reaching the English coast. The effectiveness of the self-defending bomber was now in serious doubt. Further losses during the battle for

Norway and Denmark in April convinced Bomber Command that the turret-armed bomber could no longer operate effectively in daylight. On April 12, 1940, the night-bombing campaign against Germany began. Air gunners who had trained for daylight operations now had to retrain themselves for long missions in total darkness. The gunner's role changed from active defender to sharp-eyed observer, where the best gunners were the ones who never fired a shot.

"I never fired my guns on ops. I always believed that if you fired your guns you were giving your position away, and if you could see a fighter first, it was always best to veer away from it. It was just luck, you couldn't really see that far at night.

"I only knew of one fella who shot down a Messerschmitt over Magdaberg. That's the only gunner I ever knew who shot down anybody! I was a little concerned about another fellow in the next squadron who had about five or six! I don't how many of them were Halis and Lancasters, but nobody wanted to fly anywhere near him."

JOE DAVIS, WIRELESS AIR GUNNER,
HALIFAX MID-UPPER, 429 SQUADRON, 6 GROUP, RCAF

According to *The German Night Fighter and The RAF Night Bomber Report RAF and GAF Aircrew Compare Notes A.D.I (K) Report No. 337/1945*, which relied on information from German night fighter pilots Major Schnaufer, Kommodore of NJG 4, and two of his Gruppe-kommandeure: "Of the bombers shot down by NJG 4 pilots, approximately 40 percent did not fire or manoeuvre, 50 percent fired or manoeuvred after the fighter had opened fire, and 10 to 20 percent fired or manoeuvred first."

Not all daylight operations against Germany ended. A small force of Blenheims continued daylight attacks, but they suffered heavy losses. Between July and September

In the USAAF, the safest position in a heavy bomber was the ball turret. These conclusions were drawn from a detailed analysis in which 1,117 air-battle casualties were tabulated. These figures were drawn from 8th Air Force B-17 and B-24 Bomb Groups. There were 110 men killed, 1,007 wounded. (*Many heavy bombers carried two waist gunners, so the 20.9 figure must be reduced substantially.*)

Crew Position	Killed and Wounded	Percent
Bombardiers	196	17.6
Waist gunners	233	20.9*
Tail gunners	140	12.5
Navigators	136	12.2
Radio operators	95	8.5
Top turret gunners	94	8.4
Pilot	83	7.4
Co-pilot	74	6.6
Ball turret	6	5.9

Crew Position and Survivability in Bomber Command Aircraft
January–June 1943 (Percent)

Crew Position	Lancaster	Halifax	Wellington
Pilot	9.6	20.8	14.6
Navigator	13.8	36.2	21.0
Wireless operator	11.9	32.5	18.5
Flight engineer	12.4	34.0	-
Bomb aimer	13.2	31.4	18.5
Mid-upper gunner	8.5	27.3	-
Rear gunner	8.0	23.4	14.6

From *The Crucible of War 1939–1945. The Official History of the Royal Canadian Air Force Volume III.*

1941, the RAF took delivery of twenty B-17C Flying Fortresses. At this time the B-17 was armed only with one hand-held .30-caliber machine gun in the nose and six .50-caliber guns in the dorsal, waist and ventral gun positions. The high-altitude bombing missions proved a failure. Losses due to training and operations were considerable, and the gun positions designed to defend the B-17 had proved inadequate.

The British summarized the B-17 defensive firepower thus: "It is concluded that the existing gun armament is valueless except for moral support and should be removed. Armour plate and self sealing tanks are already fitted, but the reduced gun load should allow more armour to be fitted; the 'gun bath' beneath the fuselage could also be removed, improving the performance. It is suggested that steps be taken to install hand held .303 guns in the Fortress on mountings similar to those in the Hampden. Another form of defence suitable to this aircraft is the Short Aerial Mine."

The evidence was clear. Daylight bombing operations without fighter escort were not feasible. The lessons from the First World War had been ignored. Technology had given air commanders a false hope. It was a hope that would cause the death of thousands of air gunners.

In August 1937, fifty-four of the Japanese Navy's latest bomber, the Mitsubishi G3M, were shot down in three raids over China. The Japanese quickly realized that air superiority and fighter escorts were vital for successful bomber operations, and rushed their latest fighter, the Mitsubishi A5M, to the theater. In Germany, the Luftwaffe developed the long-range Bf 110 escort fighter, though to a large extent they still believed that the speed and armament of their medium bombers would suffice. The Germans were on the right track, but the Bf 110 proved a failure. During the Battle of Britain, the German bomber force was decimated. The long-range Bf 110 was unable to cope against the RAF Spitfires and Hurricanes. Although the German bombers were provided with an escort, the short range of the Bf 109 was terribly inadequate. If the Bf 109 had been equipped with a drop tank, the Battle of Britain would have been very different. Like the British, the Germans turned to the cover of darkness to continue their bombing campaign. Now it was the Americans' turn to learn the lesson.

The British experience with the B-17 did not dissuade the Americans. It only spurred them on. They realized that the B-17C was not a combat-ready aircraft and that improvements in its defensive armament were paramount. Armor and self-sealing tanks were added. The Americans also argued that the British had used the B-17 only in twos and threes. Large formations of B-17s were required to get the full effect of interlocking fields of fire. The theoretical strength of the B-17 and B-24 depended on their heavy armament of .50-caliber machine guns. The B-17F carried twelve .50-caliber guns and the B-24D carried eleven. A combat wing of fifty-four B-17s bristled with 648 machine guns, each aircraft carrying 9,000 rounds of ammunition. The standard M2 Browning .50-caliber machine gun fired 750 rounds a minute up to an effective range of 600 yards. The two-ounce projectile remained lethal to humans up to four miles. German fighter pilots who attacked the large B-17 formations from behind flew straight into a wall of fire.

"The Americans were much more powerful with their Flying Fortresses. During the war, I made about five or six daylight missions. We had very serious losses when the night fighters were told to attack in daylight. Those American gunners started shooting as soon as they saw a little point. It was like flying into a shower. I remember I was attacking a formation of B-17s with four Bf 110 night fighters over the Baltic. We were attacking the way we were taught in Night Fighter School — from the rear. My radar operator said, 'Captain, the distance is now three kilometers,' and it made no sense for me to start shooting from that distance because I knew my 20-mm cannon had a two-kilometer range. So as I approached — three kilometers, two kilometers — and just as I was about to shoot I saw for the first time in the my life something strange happening with my left and right wings. It was as if a man was poking holes with his fingers into my wings. It was the American gunners! They had started shooting. And then bang! One shell hit my canopy. It was then I realized I was flying into a shower of bullets. When I was ready to shoot, I couldn't, my electricals were out. I lost one engine, made a steep turn and got away."

PETER SPODEN, BF 110-JU 88
NIGHT FIGHTER PILOT

On August 17, 1942, the newly formed 8th Air Force flew Mission 1. Twelve B-17s from the 97th Bomb Group attacked the Sotteville railroad marshaling yards near

TOP RIGHT: AIR GUNNERS PREPARING TO FIRE TWIN .303 MACHINE GUNS MOUNTED ON A TRIPOD AT A FIRING RANGE.
BOTTOM RIGHT: GUNNERY PRACTICE WITH .50-CALIBER MACHINE GUNS ON FREE MOUNTS AND TRUCK-MOUNTED TURRETS. AMERICAN GUNNERY SCHOOLS PRODUCED NEARLY 215,000 GUNNERS WHILE THE BRITISH COMMONWEALTH AIR TRAINING PROGRAM TRAINED 34,196.

Rouen, France. The mission was a success with no losses. Ironically, this mission was escorted by four squadrons of RAF Spitfires; two Spitfires were shot down. More missions would follow and optimism was high, but it became clear that the air gunners needed better training. Brigadier-General Ira C. Eaker, commander of the 8th Air Force, wrote: "I am absolutely convinced that the following measures are sound.... Three hundred heavy bombers can attack any target in Germany by daylight with less than four percent losses." Up until this point the Americans had not penetrated German airspace and most missions had fighter escort. The German fighters based in France and the Low Countries quickly developed attack methods to take advantage of the B-17's shortcomings. It was discovered that the nose armament in the B-17 and B-24 was weak (one to three hand-held machine guns). The Germans no longer attacked from the stern. Now they came in head-on. The closing speeds were tremendous, up to 600 mph. At that rate, the nose gunner could only fire a quick burst and hope for the best. The Germans also began to attack in small formations of up to a dozen aircraft. Although many B-17s and B-24s were being shot down, the Luftwaffe never succeeded in stopping the large American heavy bomber formations from reaching their targets.

After a year of bombing German airfields, sub pens and other tactical targets, the 8th Air Force finally turned its attention to the more strategic objectives in Germany. Missions 78, 79 and 80 dealt with German aircraft plants in Kassel, Oschersleben and Warnemunde. Each target was zealously defended. Over three days the 8th lost forty-four heavy bombers. Mission 84 would prove even costlier.

The two targets were the Messerschmitt factories at Regensburg and the ball-bearing plant at Schweinfurt. It was a bold and daring plan. Not only was it the deepest penetration yet made into Germany, it was also the largest force launched to date. It was a crucial test of the American policy of daylight precision bombing. This was a maximum effort: 376 B-17s were dispatched along with eighteen squadrons of P-47 Thunderbolts and sixteen squadrons of RAF Spitfires. Support missions were also flown by squadrons of B-25s, B-26s, Whirlwinds and Typhoons. These aircraft attacked German airfields in order to, it was hoped, disrupt the German fighter defenses. The Regensburg force (139 B-17s) would lead the way, bomb the target, and continue on to North Africa. The Schweinfurt force (222 B-17s) would follow in the hope of catching the German fighters on the ground refueling, and return to England. The English weather intervened and threw the timing into complete disarray. The Regensburg force was delayed but managed to get airborne at about 9:30 A.M. The Schweinfurt forces didn't get airborne until three hours later. The fog also disrupted fighter escort timetables, causing many squadrons to miss their rendezvous. Only one fighter group made contact, but due to their limited range they had to turn back at the German border. German fighters began their attacks before the B-17s reached Germany. The attacks were savage and relentless. The Regensburg force lost twenty-four B-17s. The Schweinfurt force, which was three hours behind, now

faced a fully alert and reinforced Luftwaffe. Almost three hundred fighters were ready and waiting. The Germans threw everything they had at the B-17s. Crews reported Bf 109s, Fw 190s, rocket-armed Bf 110s, Me 210s and Ju 88s. When it was over, sixty B-17s had been lost. Eleven more returned to England but never flew again, and a further 162 suffered lesser damage. The air gunners claimed 288 Luftwaffe fighters, almost the entire German force that fought that day. Postwar research revealed that twenty-one were lost to gunners and a further twenty-one were claimed by escort fighters. (The USAAF did not officially recognize claims made by gunners due to the obvious problems of confirmation.)

"I never claimed any fighters shot down but I had some pretty good shots at a couple of them. Whether I hit them, and I'm sure I did, there were other gunners shooting at them as well. We flew in tight box formations, and if you're shooting at a fighter there's a bunch of other guys doing the same thing. It didn't really matter who shot him down just as long as he went down."

WILLIAM J. HOWARD, B-17 WAIST GUNNER,
351ST SQUADRON, 100TH BOMB GROUP

Over 550 men had been killed or reported missing, the majority of them gunners. The raids into Germany continued with heavy losses. On September 6, 262 B-17s bombed Stuttgart, but forty-five aircraft were lost. In October, the 8th returned to Schweinfurt with another sixty B-17s lost and 138 damaged. Losses were well above ten percent of the bombing force. That rate would eventually lead to its destruction. The 8th Air Force report, *An Evaluation of Defensive Measures Taken to Protect Heavy Bombers from Loss and Damage*, stated: "On October 14th the mission to Schweinfurt, 60 aircraft were lost — 26% of those attacking. For about three months thereafter, until a long-range escort force of sufficient size was available, attacks were confined to French and nearby German targets."

"The day I arrived at the 100th Bomb Group, they went to Schweinfurt. Only one plane made it back. I think they sent out thirteen bombers that day. I have to say I was kind of concerned. We couldn't start our combat missions right away because we had to get new aircraft and the new crews had to learn evasive combat formation flying."

DON ATKINSON, B-17 TOP TURRET GUNNER,
418TH SQUADRON, 100TH BOMB GROUP

A VERY BRAVE BF 110 PILOT PRESSING HOME HIS ATTACK ON A B-17.

It was obvious that the self-defending bomber could not do the job. Without long-range escort fighters, the bombing of Germany would be a losing battle. History has claimed it was the North American P-51 Mustang that saved the daylight bomber offensive. In reality it was the lowly drop tank and the great performance of the P-51 that helped tip the scales. Equipped with two 108-gallon drop tanks, the P-51 could fly to Berlin and back.

In the night sky over Germany, RAF Bomber Command air gunners were also fighting a losing battle. Since the RAF had switched to night bombing in 1940, Germany had been subject to constant bombing raids of various strengths. Early night raids proved wildly ineffective. Most crews never found their targets, and when they did, their chances of hitting the target were minimal. The technology and techniques used at the time were clearly inadequate. But by 1943, Bomber Command was steadily growing in strength and improving with new navigation aids such as Oboe and H2S airborne ground-scanning radar. Both devices were a godsend to a force that until then had to rely on dead-reckoning navigation. The Pathfinder Force was the first to receive the new H2S radar. The Pathfinder Force was formed to provide Bomber Command with a highly trained unit capable of finding and marking targets at night. The Pathfinder crews would fly ahead of the main force and use their H2S radar to find the target city and drop Target Indicators, brightly colored flares. The main force would follow and drop their bombs on the TIs. At the beginning of 1943, Bomber Command could dispatch up to 450 heavy bombers, most of those four-engine Halifaxes, Lancasters and Stirlings.

Like the Americans, the British relied on the powered turret to defend the bomber. While the Americans equipped their aircraft with the hard-hitting, robust .50-caliber, the British retained the rifle-caliber .303-caliber Browning machine gun. The only advantage the .303 had was its rate of fire: up to 1,200 rounds per minute. A four-gun turret could pour out 4,800

rounds a minute! At short range it could be devastating, and with the close quarters required for a night fighter to shoot down a bomber at night, the four-gun turret was certainly a deterrent. Still, it must be remembered it was not the gunner's job to shoot down the enemy. His job was to keep the enemy fighter from shooting him down.

"I fired my guns in anger over Stettin one night. I took a whack at a Fw 190. There were searchlights all over the place. It appeared as though the Fw 190 was riding up this one beam. That beam must have been in front of us, because when you looked down you could see the front profile of the aircraft. Now, he was quite possibly going after a plane ahead of us following the searchlight, maybe there was one coned, I don't know. I fired a burst and immediately thought, 'forget this nonsense,' and we did a corkscrew starboard. That was scary. The skipper had problems pulling out because of the speed. The corkscrew was so violent, I swear I saw those darn rudders twisting."

FLYING OFFICER BILL COLE, AIR GUNNER HALIFAX AND LANCASTER REAR TURRET, 428 SQUADRON, 6 GROUP, RCAF

THIS UNIQUE COLOR SHOT, TAKEN FROM THE TAIL GUNNER'S POSITION, SHOWS B-17S OF THE 385TH BOMB GROUP PLOWING THROUGH DEADLY FLAK OVER THE TARGET.

According to *The German Night Fighter and The RAF Night Bomber Report*, "All pilots considered the corkscrew a most effective evasive manoeuvre, but they were of the opinion that a corkscrewing Halifax was an easier target than the Lancaster, although the Lancaster caught fire, when hit, more easily."

Most bomber crews used evasive tactics such as the corkscrew maneuver to avoid night fighters. Most air gunners did not want to fight it out with a heavily armed German night fighter, and for good reason. Most were armed with two 30-mm and four 20-mm cannon. On average, twenty hits of 20 mm were required to shoot down a four-engine heavy bomber. With the 30 mm, three or four hits were usually enough. During the war, ninety percent of the Lancaster sorties were made without fighter contact, but of the ten percent that were, half were shot down. At night, the air gunner was not the aggressor but the defender. The best defense at night was to see the enemy first, call for evasive action, and hope that the pilot responded in time. In combat, it was the gunners who effectively controlled the aircraft. Most crews never saw the attacking enemy fighter. Many bomber crews owe their lives to the alertness and quick action of their gunners.

"On the night of August 16, 1944,we were attacked by fighters on the way to Kiel. I was not expecting something like that to happen, although always a possibility. I was concentrating on my search pattern. Lost in my thoughts, I continued my search pattern when I was suddenly frightened out of my wits. The rear gunner screamed into the intercom, 'Starboard go!' and immediately the aircraft went into a violent evasive action. As the plane dove down, I had no idea where the fighter was and became quite disoriented, momentarily not knowing where up and down was. I believe at that instant I was perhaps terrified and I could hear the tat, tat, tat of the rear turret guns in my headphones. I didn't know where the hell he was firing or where the fighter was."

FLYING OFFICER DOUG SAMPLE, HALIFAX AIR GUNNER, 415 SQUADRON, 6 GROUP, RCAF

While the Americans were learning a bitter lesson during the day, the British were suffering heavy losses at night. By the end of 1943, the German night-fighter force had grown in numbers and sophistication. Many units were equipped with the Bf 110 twin-engine fighter. In the

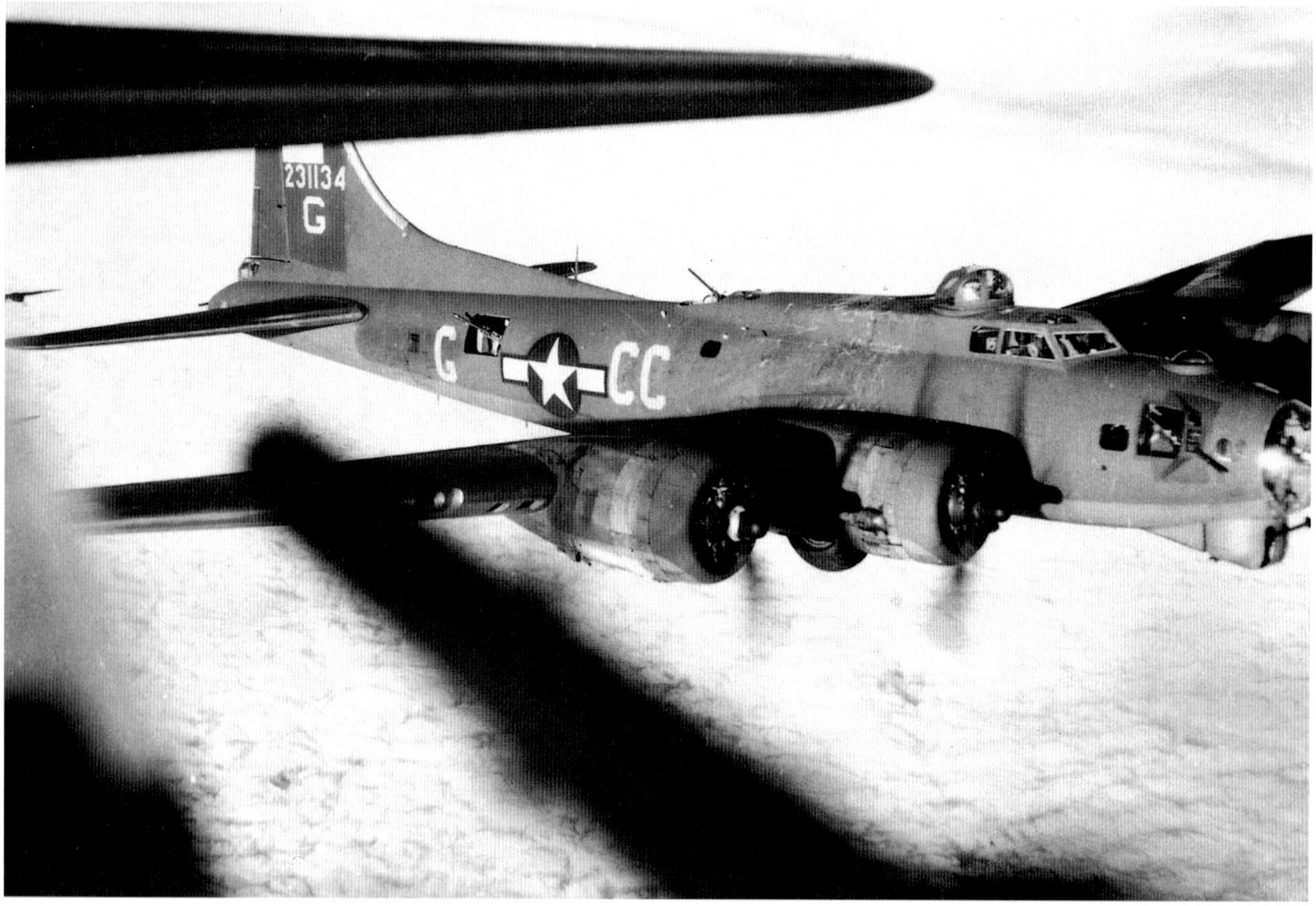
231134
G
G
CC

AMERICAN DAYLIGHT BOMBING TACTICS REQUIRED TIGHT FORMATIONS. THIS PHOTOGRAPH, TAKEN FROM THE WAIST POSITION OF A B-17, SHOWS HOW CLOSE THE BOMBERS ACTUALLY FLEW TO EACH OTHER.

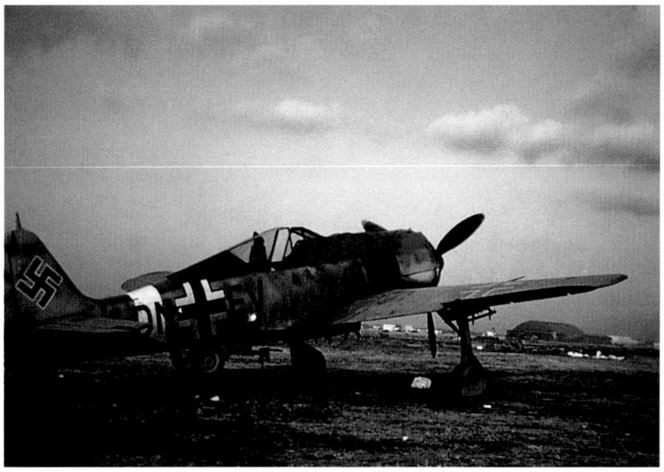

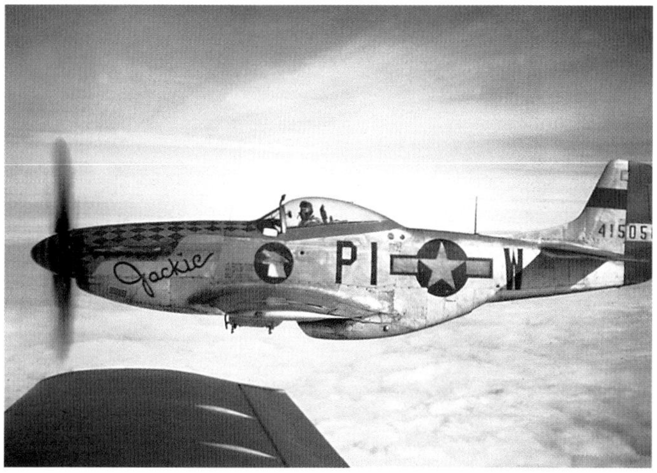

Jackie
PI
W
41505

LEFT: AN FW 190 G-3, A FIGHTER BOMBER VERSION OF THE FAMOUS FIGHTER. TO COUNTER THE AMERICAN HEAVY BOMBER RAIDS, A SPECIAL *STRUMGURPPE* OF FW 190S WAS FORMED. THESE FORMATIONS WERE EQUIPPED WITH THE HEAVILY ARMED FW 190A-8S. ARMAMENT CONSISTED OF TWO 30-MM CANNON, TWO 20-MM CANNON AND TWO 13-MM MACHINE GUNS.

RIGHT: INBOUND TO ENGLAND FROM GERMANY, THIS 365TH FIGHTER GROUP P-51 HAS JUST COMPLETED ANOTHER LONG-RANGE ESCORT MISSION, APRIL 1945.

beginning, the Bf 110 filled the role admirably, but it soon lost its superior position in the German night-fighter force. The additional weight of radar, exhaust covers and a three-man crew robbed the comparatively light 110 of vital speed. Its speed was so degraded that a Lancaster free of its bomb load was very hard to catch. The Junkers Ju 88, however, was fast and powerful, with two hours flying endurance. The Luftwaffe started re-equipping the 110 units with the Ju 88, but the conversion went slowly. The other threat that faced the air gunners, one they could do nothing about, was flak. A Bomber Command report written in 1944 shows there were an estimated 20,625 anti-aircraft guns and 6,880 searchlights deployed in Germany and the West. Flak was a very inefficient means to shoot down bombers, but it did force them to fly higher. This caused a loss of bombing accuracy and limited the routes available to the Bomber Command planners, which in turn gave the advantage to the German fighter-controllers.

As the Americans straggled home from their bruising raids on Schweinfurt and Regensburg, RAF Bomber Command was preparing for a night raid on Peenemunde on the Baltic coast. This raid was designed to halt or at least set back the development of the German V-2 rocket program. The raid was carried out in moonlight to increase the chances of success. In all, there were 596 bombers — 324 Lancasters, 218 Halifaxes, 54 Stirlings and 8 Mosquitos — taking part. The Pathfinders found Peenemunde without difficulty. A Mosquito diversion aimed at Berlin drew off most of the German night fighters, but that would not last. As the last wave of RAF bombers began their attack, the German night fighters realized their mistake and raced toward Peenemunde. Forty aircraft were shot down — 23 Lancasters, 15 Halifaxes and 2 Stirlings — 6.7 percent of the force dispatched. If the Germans had not been diverted, RAF losses would have been much higher. This was also the first night the Germans introduced their new *schräge Musik* weapons; these twin upward-firing cannon were fitted in the rear gunner's cockpit of the Bf 110. It is believed that two

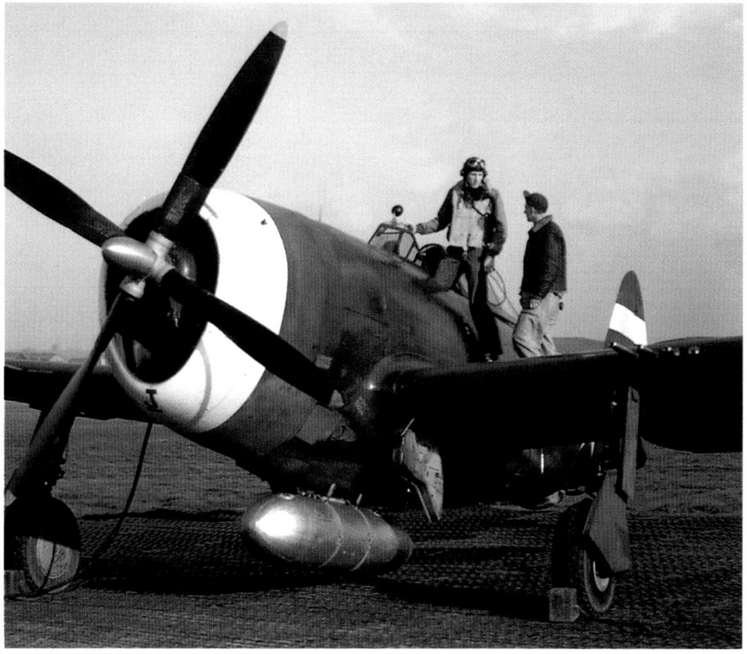

THE 108-GALLON PAPER DROP TANK EXTENDED THE RANGE OF THE P-47 AND P-51. THIS P-47 CARRIES JUST ONE TANK. A BRITISH INVENTION, THE LAMINATED AND GLUED PAPER TANKS COULD HOLD FUEL FOR JUST ONE MISSION. AFTER A FEW HOURS THE GLUES WOULD BREAK DOWN AND THE TANKS WOULD DISINTEGRATE.

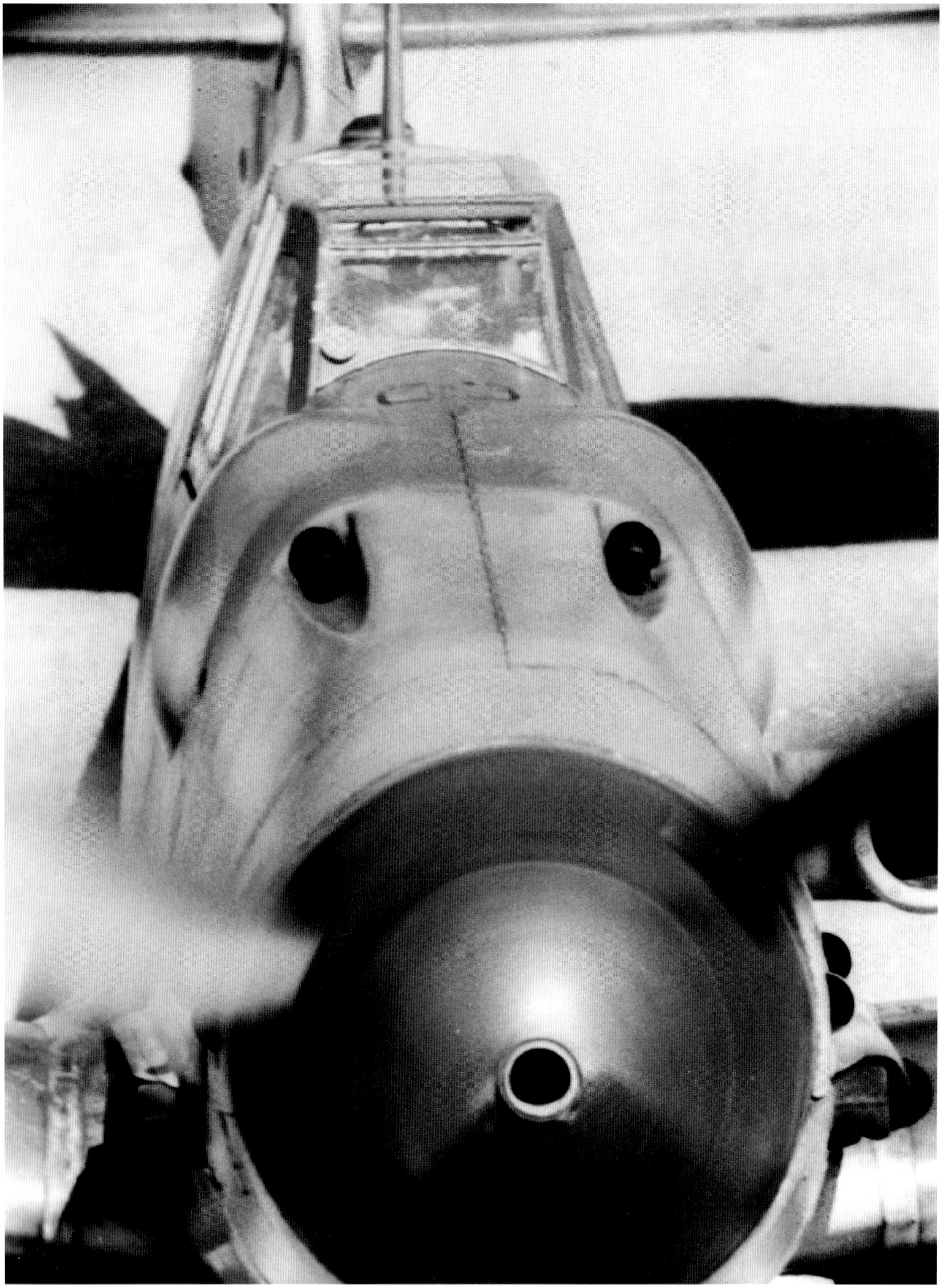

The sharp end of the Bf 109. Later versions of the G model were armed with two 13-mm machine guns over the engine and one 30-mm cannon firing through the spinner.

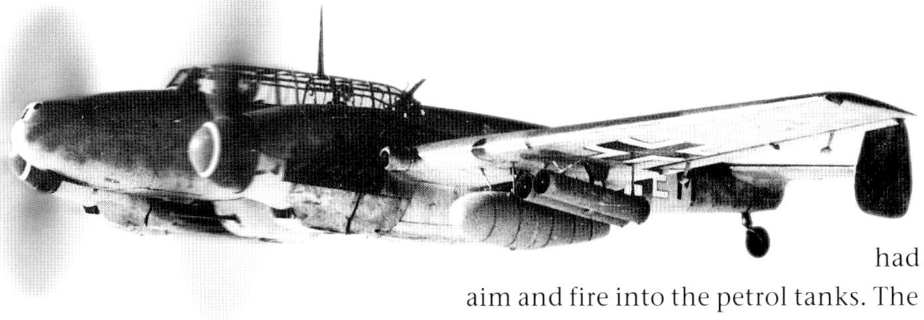

schräge Musik aircraft shot down six of the bombers lost on the raid. On August 23, Bomber Command returned to Berlin with 727 aircraft. Fifty-six bombers were lost, 7.9 percent of the heavy bomber force. It was Bomber Command's greatest loss of aircraft in one night so far in the war.

The upward-firing cannon beginning to equip more and more German night fighters served as an extremely effective weapon, and the RAF seemed to know little or nothing about it. The Germans effectively exploited the bomber's great weakness — no mid-under gun turret.

> "We were told at the beginning that the Lancaster and Halifax did not have a ventral turret. I guess it was because of the way they were constructed."
>
> PETER SPODEN, BF 110 AND JU 88 NIGHT FIGHTER PILOT

The standard method of attack used by the Germans involved a radar approach from behind. Once a visual sighting was made, the German pilot approached the unsuspecting bomber from below and behind. When the correct position and range were reached, the fighter, in a slightly nose-up attitude, would fire its forward-firing guns at the bomber's fuselage or at the wing between the fuselage and engines. This approach exposed the fighter to the rear turret, but with *schräge Musik*, the fighter could approach the bomber from below and well out of range of the rear turret guns. From there, the fighter entered the bomber's blind spot. Most of the bombers in the RAF had no belly guns to protect them from attack from below. Once in this position, all the German pilot had to do was take aim and fire into the petrol tanks. The upward-firing cannon were extremely effective, so successful that few British aircraft managed to return to report on such an attack. Some British bombers were equipped with a ventral turret, but ironically, most of these were removed just as the upward-firing cannon began to appear. It seems, but cannot be confirmed, that 6 Group (Canadian) had an inkling of the weapon. They were fitting their aircraft with mid-under guns just as fast as they could! It is not known how effective these modifications were.

> "When we received special training as mid-under gunners with .50-caliber Brownings, we were told that the Germans were attacking bombers from underneath. At the time, we were not told anything about *schräge Musik* [slanted vertical-firing 20-mm cannon], and I only learned about it after the war. We did know that their tactics were to get vectored into the bomber stream at night and fly along until they encountered turbulence from a bomber. Once they made visual contact they would drop down low, swing underneath, and then fire into the wing tanks, setting the bomber on fire before diving away. Because of this 'new' tactic, our job as mid-unders was to watch for enemy fighters that might try and creep up underneath us."
>
> FLYING OFFICER DOUG SAMPLE, HALIFAX AIR GUNNER, 415 SQUADRON, 6 GROUP, RCAF

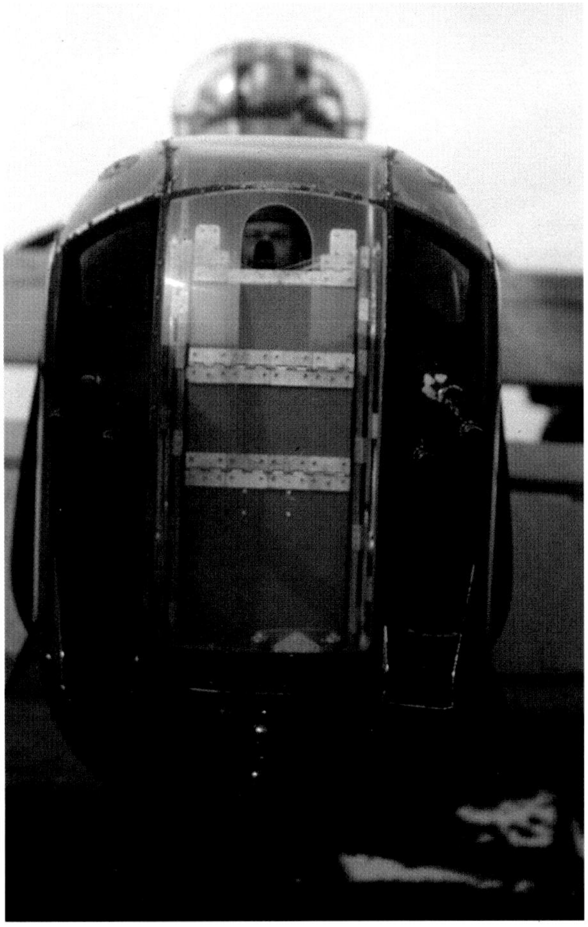

TOP: A BOMBER DESTROYER. ARMED WITH FOUR WGR 210-MM AIR-TO-AIR ROCKET-PROPELLED MORTAR SHELLS, 20-MM AND 30-MM CANNON, THE BF 110 WAS ARGUABLY THE MOST HEAVILY ARMED FIGHTER OF THE WAR.

BOTTOM: WORLD WAR II COLOR PHOTOGRAPH SHOWS THE FRAZER NASH FOUR-GUN ARMAMENT AND ARMOR PLATING DESIGNED TO HELP PROTECT THE GUNNER. THE ARTICULATED ARMOR SCREEN WAS FITTED TO THE FN 20 BECAUSE OF HEAVY CASUALITIES.

In early 1944, the German night fighter force comprised about 15 percent of the Luftwaffe. When Bomber Command set out for Nuremberg on March 30, the night fighter force was 361 twin-engine aircraft with about 150 single-seat fighters. There were 795 bombers dispatched. *The RAF Official History* called it a "curious operation," marked by "unusually bad luck and uncharacteristically bad and unimaginative planning." The raid took place under a half moon; normally this condition was considered too bright for such an attack, but Nuremberg was a lush target. Weather also played a factor. It was thought that tail winds and high cloud would help the bombers en route. But as crews learned the details of the mission, they were horrified by the prospect of flying on a moonlit night and the long, straight flight into the target. This approach put them near the Ruhur defenses and right past German night-fighter radio beacons. The Germans ignored the diversion raids and concentrated on the main force. Strong winds dispersed the bombers; the forecast high cloud did not materialize; and most of the bombers left condensation trails, making them stand out in the night sky. Over two hundred night fighters tore into the bomber stream. In the running battle that followed, ninety-five bombers were hacked from the sky, ten crashed in England and were written off, with seventy more damaged. (If the Germans had delayed their interception takeoff time by twenty minutes, the British casualty list would have been lower.) Nine German aircraft were claimed destroyed by the gunners. It was a disaster and the biggest Bomber Command loss of the war.

The bomber war in the Pacific was one of extreme ranges. The ability to project air power over long distances was vital for victory for both the Japanese and the Americans. In the development of their grand strategy for victory in the Pacific, the Japanese had built an effective long-range bomber force. They also built a long-range fighter to protect their bombers, something the Americans and British failed to do. At the beginning of the war in the Pacific, the Zero was able to escort bombers from Formosa to Manila in the Philippines, a radius of action of about 500 miles. But range came at a price. Mitsubishi's G3M Type 96 bomber had a range of 2,000 miles, but it lacked armor for its engines, fuel tanks and crew. Like all bombers, it was vulnerable to fighter attack and suffered accordingly. Japanese bombers were also equipped with hand-held machine guns and gunners for self-defense. Most of the Japanese weapons were the Army Type 89, 7.7-mm gas-operated machine gun. The Army Air Force also used a license-built copy of the German MG15 7.92-mm along with the Type 1 Ho-103 12.7-mm and Type 97 Ho-1 20-mm cannon. The Navy's standard weapon was the Type 92, 7.7-mm (.303) machine gun, derived from the Lewis gun. The Japanese, along with the French, Italian and Russians, were the only countries to equip their bombers with 20-mm cannon.

The Japanese surprise attack at Pearl Harbor at 7:55 A.M. on December 7, 1941, pushed America into global war. Hours later and still unprepared, the American forces in the Philippines suffered another surprise attack. In a matter of days, 165 aircraft out of a force of 235 were destroyed. Of the thirty-five B-17Cs located in the Philippines, only fourteen found refuge in Australia.

Like their German counterparts, the Japanese learned to attack the B-17 and B-24 from the front. But the Japanese fighters were not as heavily armed as the Bf 109s and Fw 190s. The Zero's light armament of two 7.7-mm machine guns and two 20-mm cannon proved ineffective against the B-17 and B-24. The Zero also lacked armor plate and self-sealing tanks. Against a well-flown formation of B-17s, the Zero was extremely vulnerable. In 1942, the U.S. Army Air Force decided to phase the B-17 out of the Pacific and replace it with the longer-legged B-24. At peak strength, there were eleven and a half B-24 Bombardment Groups in the Pacific. By mid-1942, USAAF crews were reporting fewer interceptions and noted that

Possibly the only image of a German night fighter (Ju 88) in pursuit of an RAF bomber over Hamburg, April 8/9, 1945.

the Japanese pilots seemed inferior compared to those encountered earlier, but they also reported that the Japanese were using a new tactic. The British and Germans had tried air-to-air bombing early on, but found the technique difficult and inaccurate. When it did work, a well-placed bomb could do more damage than anti-aircraft fire, but getting the bomb to explode at the right height and in the right place was largely guesswork. But the Japanese persisted and began to develop their air-to-air bombing techniques, and by late 1944 their attacks were effective and deadly. Against the B-24s, they developed a very efficient attack method. Using a formation of five fighters, the Japanese would attack from 12 o'clock high in trail formation. The first two fighters would drop their phosphorous air-to-air bombs and break to the left and right. The remaining three fighters timed their dives to arrive when the bombs went off. They came in head on, guns blazing.

The gunner's war in the Pacific was very different from the war their European counterparts were fighting. For the most part, the B-24 did not fly in large formations or at high altitude. Most of the targets were tactical in nature — shipping, navy bases, staging areas and airfields. The opposition was lighter, and crews did not have to fly through belts of anti-aircraft guns or thick fighter defenses. According to a report of the Statistical Control Division, *Japanese Aircraft Versus U.S. Aircraft in Aerial Combat Southwest Pacific Theatre June 5, 1943*, air gunners in the Pacific racked up an impressive score. "The heavy bombers had an extraordinary record, shooting down 180 or 13.7% of the 1,318 Japanese planes which intercepted them. Our medium bombers, on the other hand, shot down 18, 10.1% of the 178 intercepting Japanese fighters. Of particular interest is the comparison of the performance of the B-17 with that of the B-24. It seems clear that the B-24 has been fully as effective, if not more so, than the B-17 in this theatre. It has shot down 18.4% of the intercepting enemy planes compared with 10.1% by the B-17."

Liberator crews also flew night missions, and when large missions were put on, it was not unusual for B-25 medium bombers to join the B-24s. One large strike against the Japanese stronghold of Rabaul in October 1943 consisted of 63 Liberators and 107 B-25s, escorted by 107 Lightnings. The introduction of the Boeing B-29 in 1944 put the focus on the strategic stage. The B-29 was a marvel of technology and design. It was the world's first pressurized bomber, equipped with the most advanced, computer-aided remote turret gun system. For the gunners, this sophisticated computer aiming system, capable of automatically correcting for range, altitude, air speed and temperature, made their jobs a whole lot easier. Any gunner except the tail gunner could fire more than one turret at a time. On June 5, 1944, the first B-29 raid of the war took place. Ninety-eight B-29s took off from their bases in India and attacked the Makasan rail yards in Bangkok. It was not a good start. Fourteen aircraft aborted with engine failure and only eighteen bombs fell in the target area. Five aircraft crashed on landing and forty-two diverted to auxiliary fields because their fuel ran out. By the end of the year, B-29 operations from India had ceased. Their focus now switched to the bomb groups based on the islands of Guam, Tinian and Saipan. From these islands, the B-29 was just 1,500 miles from Japan. It was surprising, even after the famous Doolittle Raid of 1942 (April 1942, sixteen B-25s flew from the deck of the aircraft carrier USS *Hornet* and bombed Japan), that the Japanese still did not take the air defense of their homeland seriously.

But by May 1944, the Japanese High Command finally accepted the inevitable — American bombing attacks would only increase and the air defense of Japan had to be strengthened. Even with their change of attitude, the

A JAPANESE KI-45 TORYU TWIN-ENGINE FIGHTER PLUNGES THROUGH A FORMATION OF B-29S AND BARELY MISSES DESTRUCTION. ONE OF THE B-29S HAS BEEN DAMAGED AND IS BEGINNING TO BREAK FORMATION.

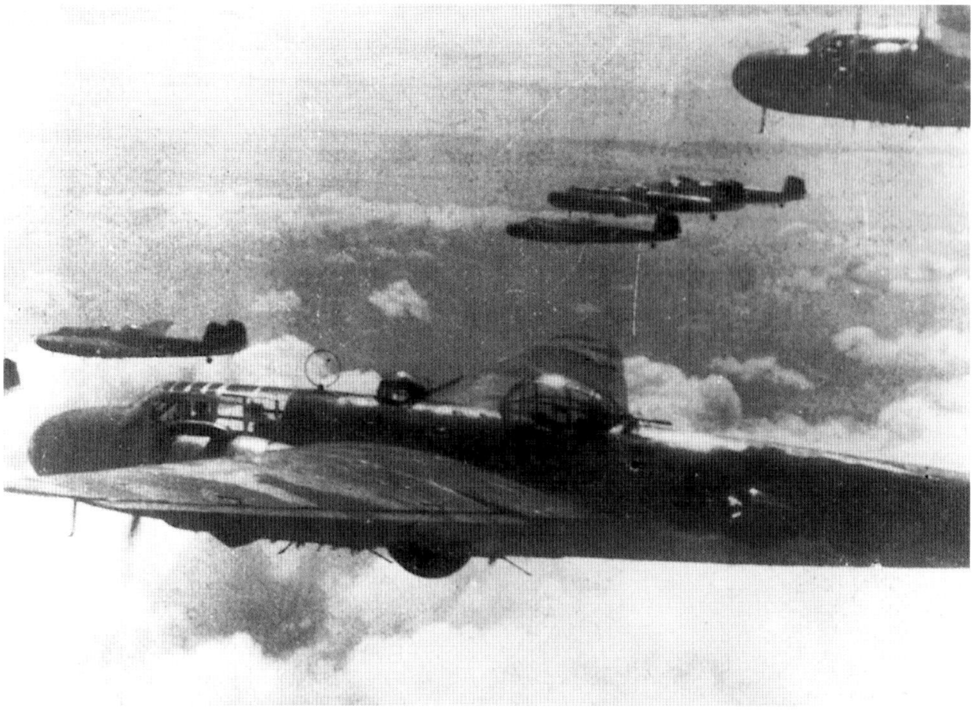

A FORMATION OF MITSUBISHI G3M3 WITH DORSAL TURRETS EXTENDED.
THE LARGE DORSAL CANOPY LOCATED RIGHT BEHIND THE TURRET WAS EQUIPPED WITH A 20-MM CANNON.

army and navy air units assigned to defend Japan were notoriously weak. Most of the fighters were old veterans: the Ki43 (Oscar), Ki44 Shoki (Tojo), with a few of the newer Ki84 Hayate (Frank), Ki45 Toryu (Nick), Ki61 Hien (Tony), and the remarkable Ki100. The navy, which controlled its own fighter forces, contributed the Kawanishi N1K1-J Shiden, the Mitsubishi A6M Reisen (Zeke) and the J2M Raiden (Jack). Most of these fighters lacked high-altitude performance and had great difficulty getting an altitude advantage on the bombers. Pilot training at this time was also extremely poor. Most if not all of the new pilots received an absolute minimum of training. Postwar analysis revealed that, compared to the German defenses, the Japanese fighter opposition was light.

> "My first mission was to Nagoya and the Mitsubishi Aircraft Factory. What I remember most was the anti-aircraft fire and one solitary fighter shooting through our formation. I remember looking across at him and seeing his head swiveling around as he shot through the formation. After he passed, I fired a short burst with no results. Apparently, someone did get him, because he bailed out."
>
> SERGEANT ANDY DOTY, B-29 TAIL GUNNER,
> 19TH BOMB GROUP

THE JAPANESE ZERO SERVED THROUGHOUT THE SECOND WORLD WAR.
BY 1944 IT WAS COMPLETELY OUTCLASSED AND INEFFECTIVE AGAINST THE HIGH-FLYING B-29.

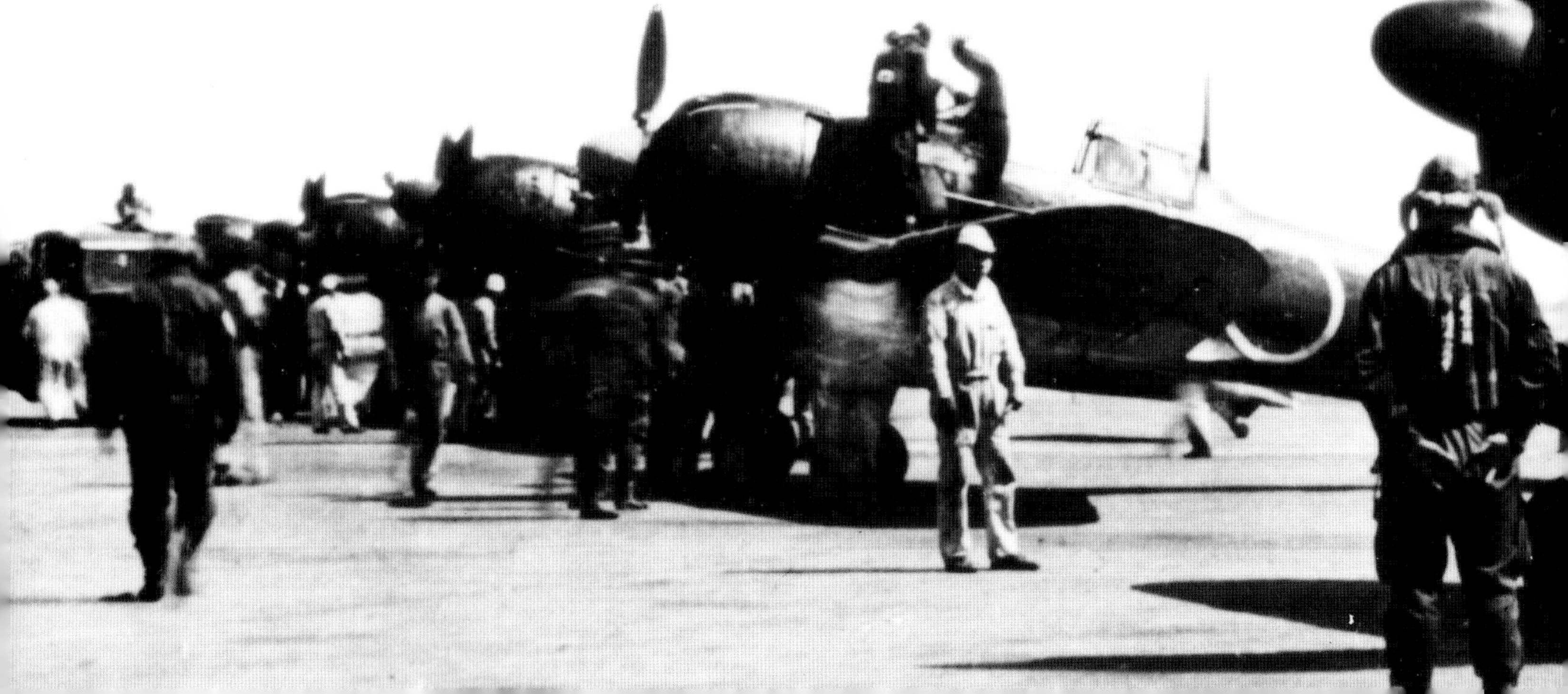

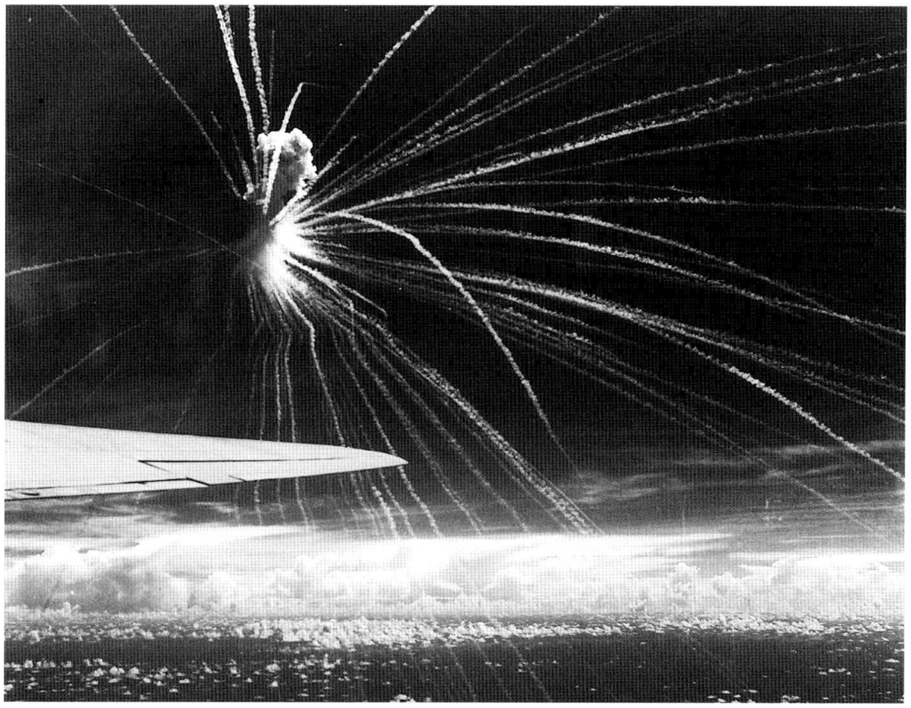

A PHOSPHORUS BOMB EXPLODES NEAR A FORMATION OF CONSOLIDATED B-24 LIBERATORS EN ROUTE TO BOMB ENEMY INSTALLATIONS ON TRUK ISLAND.

Some of the raids passed through Japanese air space completely unopposed. In looking at the claims made by B-29 gunners and the number of bombers that fell victim to fighters, it is clear how weak the Japanese opposition was. Still, as limited as the fighter opposition turned out to be, 494 B-29s were lost from all causes — AA guns, fighters, and mid-air collisions. The gunners claimed 714 fighters destroyed, 456 probably destroyed and 770 damaged. When compared to the Europe figures, these numbers seem small. But considering the fact that only 26 percent of the total Japanese fighter force was assigned to home defense, the results are startling. It must be also noted that later the B-29 raids were provided with long-range fighter escort and many were conducted at night, when the Japanese night-fighter force was weak to nonexistent.

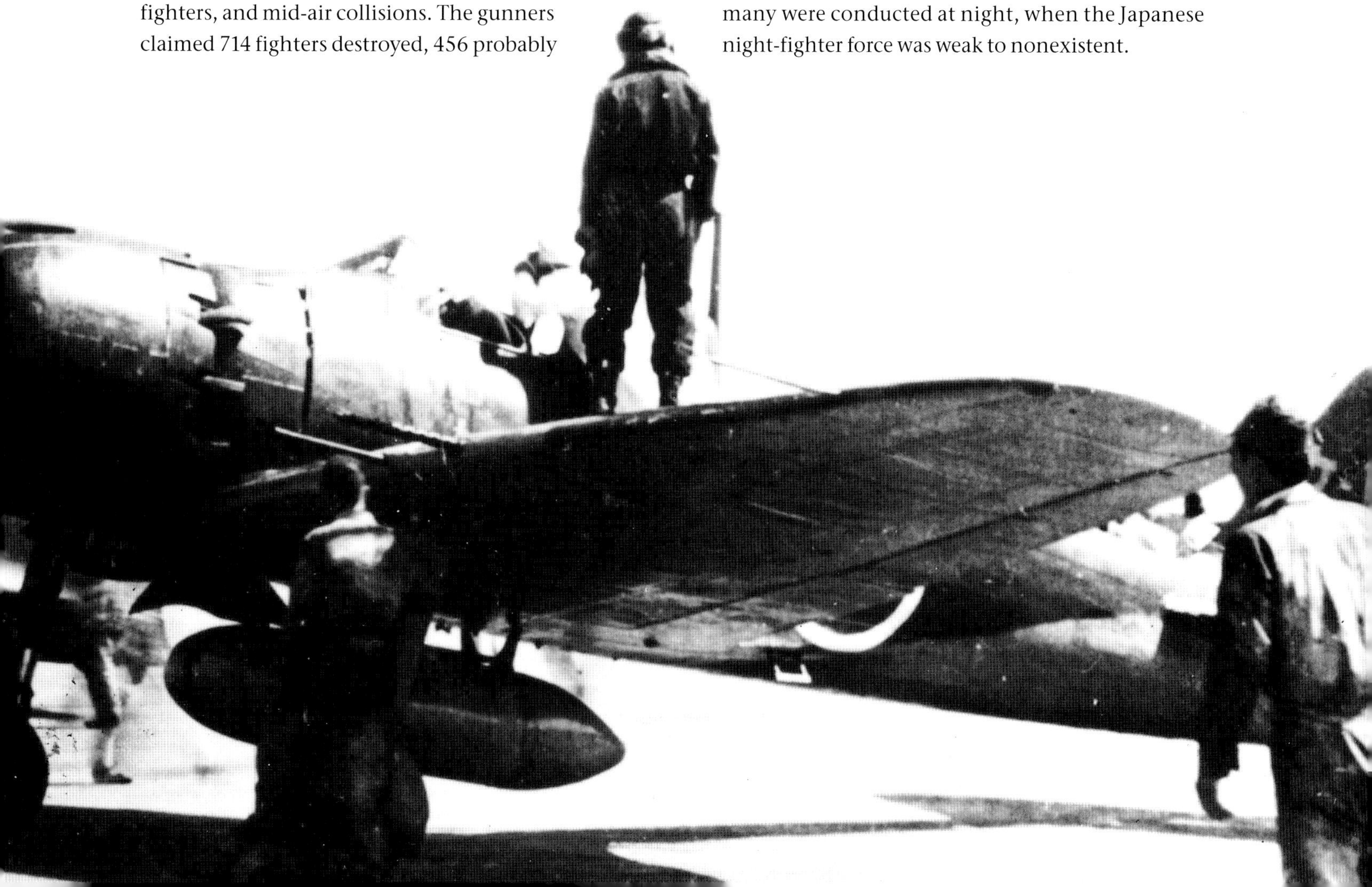

The self-defending bomber never did prove itself. No matter how brave, well trained and equipped its gunners were, the bomber was no match for fast, nimble, heavily armed fighters. Its only salvation lay with the escort fighter. While the Luftwaffe saw the need for a long-range escort fighter and developed the Bf 110, the Allies never considered such a fighter. Shortly after the war began, the British switched to night bombing in the hope that darkness would protect their bombers. It never worked, and they paid a heavy price: 7,953 aircraft were lost at night and 1,000 in daylight operations; 55,500 aircrew lost their lives, over 18,000 of them air gunners. Switching to night operations brought with it a host of new and difficult problems. Finding a factory or refinery at night was next to impossible. The British quickly realized their limitations and switched to area bombing. Finding a large German city was easier than locating a single factory.

As the war progressed and the German night-fighter defenses increased, the RAF spent an incredible amount of time and energy developing new navigational devices, along with a host of electronic countermeasures. Instead of protecting the bombers, some of these devices actually helped the Germans. The German Naxos system homed in on H2S radar; and Monica tail-warning radar, a system designed to warn the tail gunner of an approaching night fighter, had too short a range. Even worse, the Germans developed Flensburg to home in on Monica's signals!

It is often believed that the British were not interested in developing a long-range escort fighter. This is not entirely true. Even though the British authorities showed little or no interest in developing a long-range escort fighter, they did try to extend the range of the Spitfire. In August 1940, a small number of Spitfire Mk IIs were modified with a wing-mounted fuel tank. This tank, which was not jettisonable, carried an extra 40 gallons. Three squadrons, 66, 118 and 152, were equipped with this version and were involved in long-range escort for British day bombers. The British also developed the under-fuselage slipper tank. This extended the range of the Spitfire greatly, but the slipper tank was always in short supply. British ace John Johnson later wrote that "with a little foresight, Spitfires could have fought well inside Germany and could have helped the Eighth in their great venture." Although the Mk VIII Spitfire, which appeared in August 1943, had additional fuel tanks in its wings and a longer range, it was never used in Northwest Europe. Instead, it was assigned to the Mid-East, Mediterranean and Pacific theaters.

The USAAF's air gunners and their British counterparts were the best equipped in the world. The M2 .50-caliber and Browning .303-caliber machine guns were proven and effective weapons. The powered turrets, built by Sperry, Martin, Bendix, Bell, Erco, Emerson, Frazer Nash, Boulton Paul, Bristol and Consolidated, were remarkable pieces of machinery. But they were not enough to stop the heavy losses suffered by the USAAF and RAF. It was the belief in the self-defending bomber, politics and the pressure of war that sent so many young gunners to their deaths. One of the biggest blunders in the strategic bombing of Germany was the Allies' failure to develop long-range escort fighters and to build and equip its existing fighters (P-47, P-38, Spitfire) with long-range fuel tanks sooner.

The report *An Evaluation of Defensive Measures Taken to Protect Heavy Bombers from Loss and Damage* stated: "An analysis of 38 operations in July, August and September 1943 indicated that when a given number of enemy fighters attacked an unescorted formation, the losses were 7 times as great, and the damage 2½ times as great

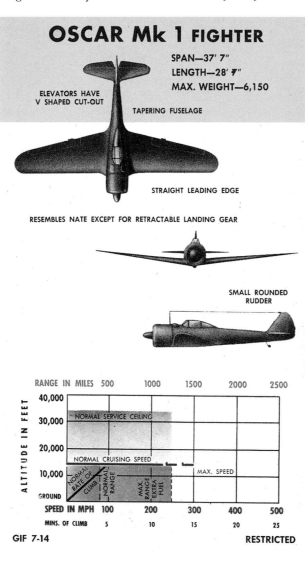

OSCAR Mk 1 FIGHTER
SPAN—37′ 7″
LENGTH—28′
MAX. WEIGHT—6,150
ELEVATORS HAVE V SHAPED CUT-OUT
TAPERING FUSELAGE
STRAIGHT LEADING EDGE
RESEMBLES NATE EXCEPT FOR RETRACTABLE LANDING GEAR
SMALL ROUNDED RUDDER
RANGE IN MILES 500 1000 1500 2000 2500
ALTITUDE IN FEET
40,000
30,000
20,000
10,000
GROUND
NORMAL SERVICE CEILING
NORMAL CRUISING SPEED
MAX. SPEED
NORMAL RATE OF CLIMB
NORMAL RANGE
MAX. RANGE EXTRA FUEL
SPEED IN MPH 100 200 300 400 500
MINS. OF CLIMB 5 10 15 20 25
GIF 7-14
RESTRICTED

AIRCRAFT RECOGNITION CHART OF NAKAJIMA KI-43 HAYABUSA, TAKEN FROM A B-29 GUNNER'S MANUAL.

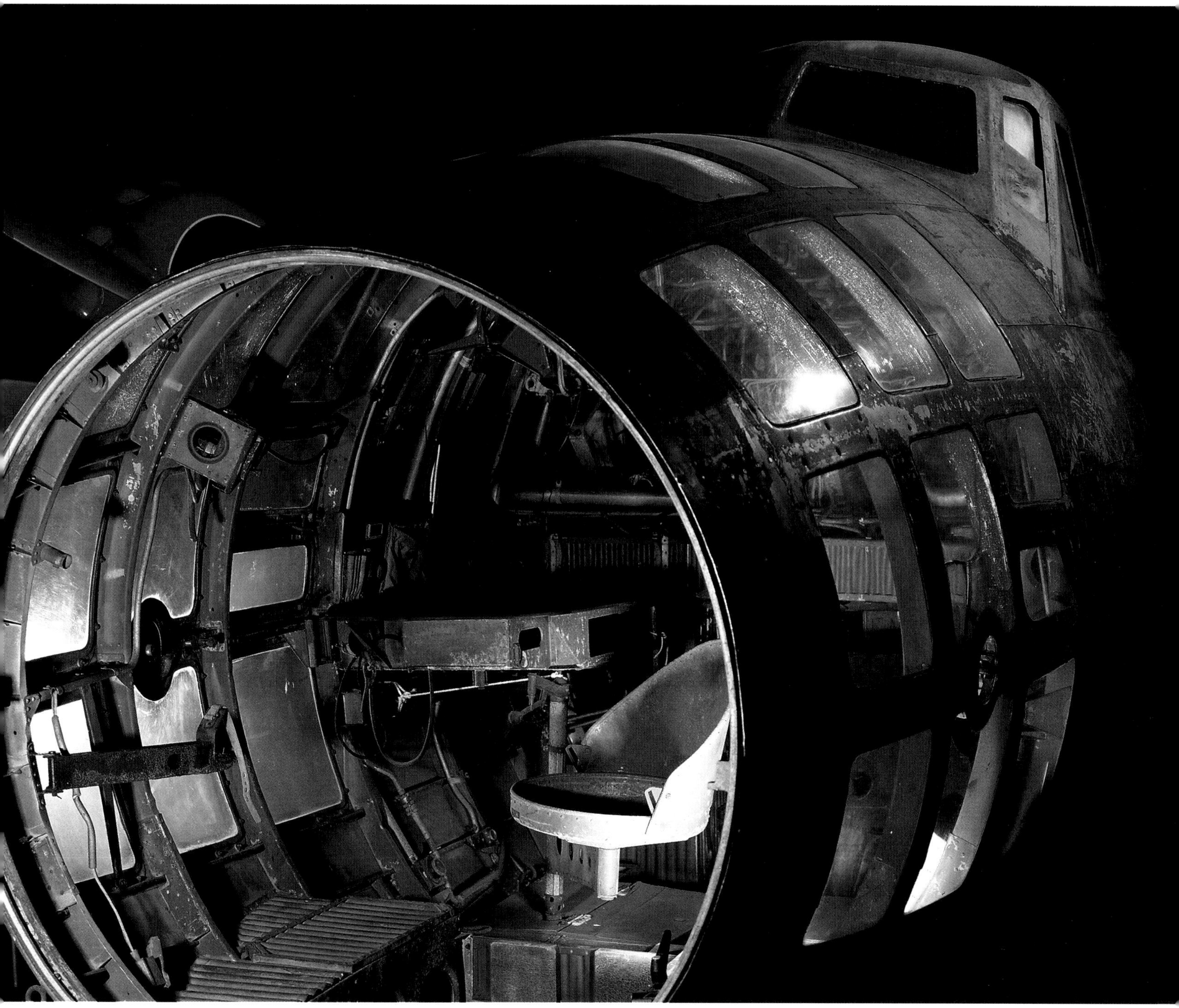

as when the same number of enemy fighters attacked a formation that had escort all the way.

"Our escort scared off or fought off much of the enemy opposition that would have taken its toll of our bombers, disrupted our formations and interfered with our bombing accuracy. They piled up an impressive record of enemy aircraft destroyed or damaged. Their efforts in the main are responsible for cutting the loss rate to fighters to a fraction of what it was, and thus saving thousands of bombers."

When the P-47 was finally equipped with the British-made 108-gallon paper tank, it had an escort radius of 375 miles. Equipped with three large drop tanks, UK-based P-47s could fly to Berlin and back. The USAAF also failed to push the modification of the P-38 sooner. Early models lacked high-altitude performance, but later versions, with better engines and improved cockpit heating, proved to be good escort fighters. The USAAF also showed a remarkable disinterest in the British-ordered P-51 Mustang. This fighter could have been in action six months earlier than

The nose section of the Mitsubishi G4M medium bomber. Defensive armament consisted of three 7.7-mm machine guns in the nose, dorsal and ventral positions and a 20-mm cannon in the tail. The G4M was the first Japanese bomber to be equipped with a powered turret.
National Air and Space Museum.

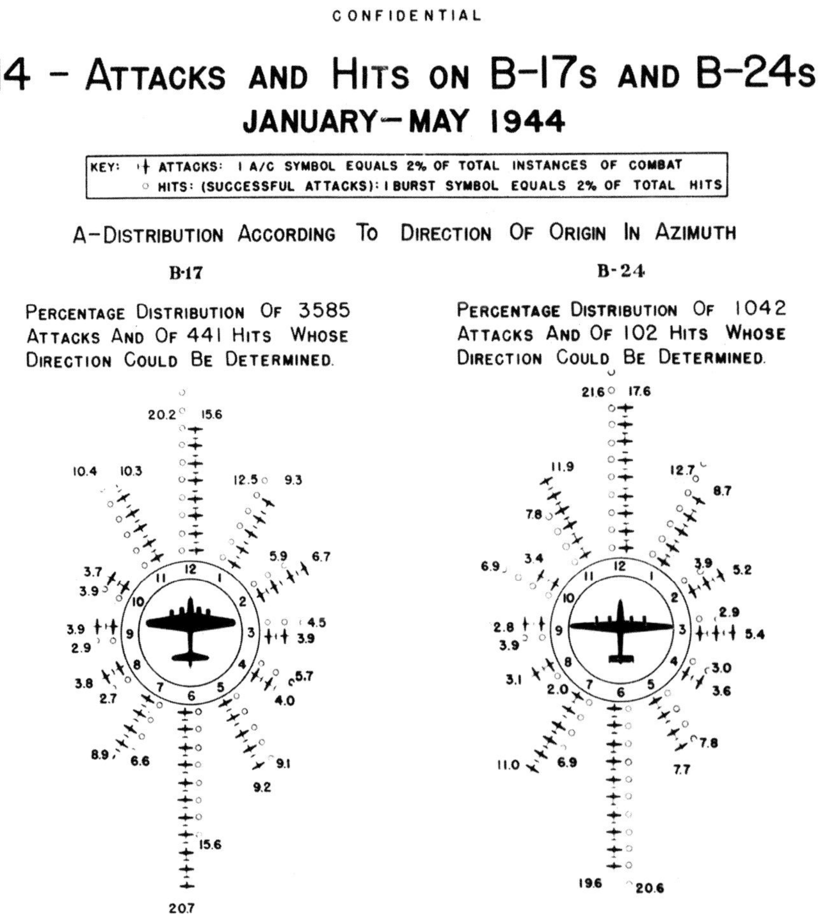

14 – ATTACKS AND HITS ON B-17s AND B-24s
JANUARY–MAY 1944

KEY: ┼ ATTACKS: I A/C SYMBOL EQUALS 2% OF TOTAL INSTANCES OF COMBAT
○ HITS: (SUCCESSFUL ATTACKS): I BURST SYMBOL EQUALS 2% OF TOTAL HITS

A–DISTRIBUTION ACCORDING TO DIRECTION OF ORIGIN IN AZIMUTH

B-17

PERCENTAGE DISTRIBUTION OF 3585 ATTACKS AND OF 441 HITS WHOSE DIRECTION COULD BE DETERMINED.

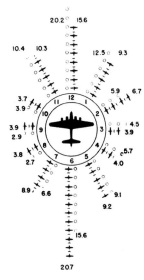

B-24

PERCENTAGE DISTRIBUTION OF 1042 ATTACKS AND OF 102 HITS WHOSE DIRECTION COULD BE DETERMINED.

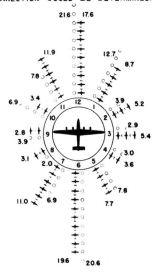

it was. Instead, it was not until December 1943 that the Merlin-powered Mustang flew their first escort missions into Germany. Equipped with two 108-gallon drop tanks, the Mustang had a range of 850 miles, making missions to any part of Germany possible. As good as the Mustang was, it alone did not save the 8th Air Force from defeat, as is sometimes supposed. Even if the Mustang had never been built, the improvements to the P-38 and the extension of the P-47's range would probably have garnered the same result. In fact, the P-47 Thunderbolt destroyed more Luftwaffe aircraft than the celebrated P-51 Mustang. William E. Kepner, Commander, 8th Fighter Command, is quoted as saying: "If it can be said that the P-38s struck the Luftwaffe in its vitals and the P-51s [gave] the coup de grace, it was the Thunderbolt that broke its back."

Too many gunners lost their lives in battles high above Germany. The failure of their leaders to see the limits of the self-defending bomber and to develop the long-range fighter sooner tragically sealed their fate. Their battle was fought against incredible odds, but in the end the thousands of young men who manned their turrets and guns and fought against frostbite, anoxia, deadly flak and the swarms of enemy fighters played a vital role in the destruction of the German Luftwaffe and Japanese Air Forces. Their contribution cannot be understated. They paid a heavy price and, in many ways, they were the foot soldiers of the sky.

ABOVE: THIS GRAPHIC, PRODUCED IN 1944, SHOWS THAT THE GREAT MAJORITY OF FIGHTER ATTACKS ON AMERICAN HEAVY BOMBERS CAME FROM THE 12 AND 6 O'CLOCK POSITIONS.

TOP RIGHT: ONE GUNNER COULD CONTROL MULTIPLE TURRETS WITH THE GENERAL ELECTRIC REMOTE-CONTROLLED TURRET SYSTEM IN THE B-29.

BOTTOM RIGHT: LONG-RANGE FIGHTER ESCORT MADE DAYLIGHT STRATEGIC BOMBING POSSIBLE. HERE A LONE P-47 PEELS AWAY FROM A GROUP OF B-17Gs FROM THE 381ST BOMB GROUP.

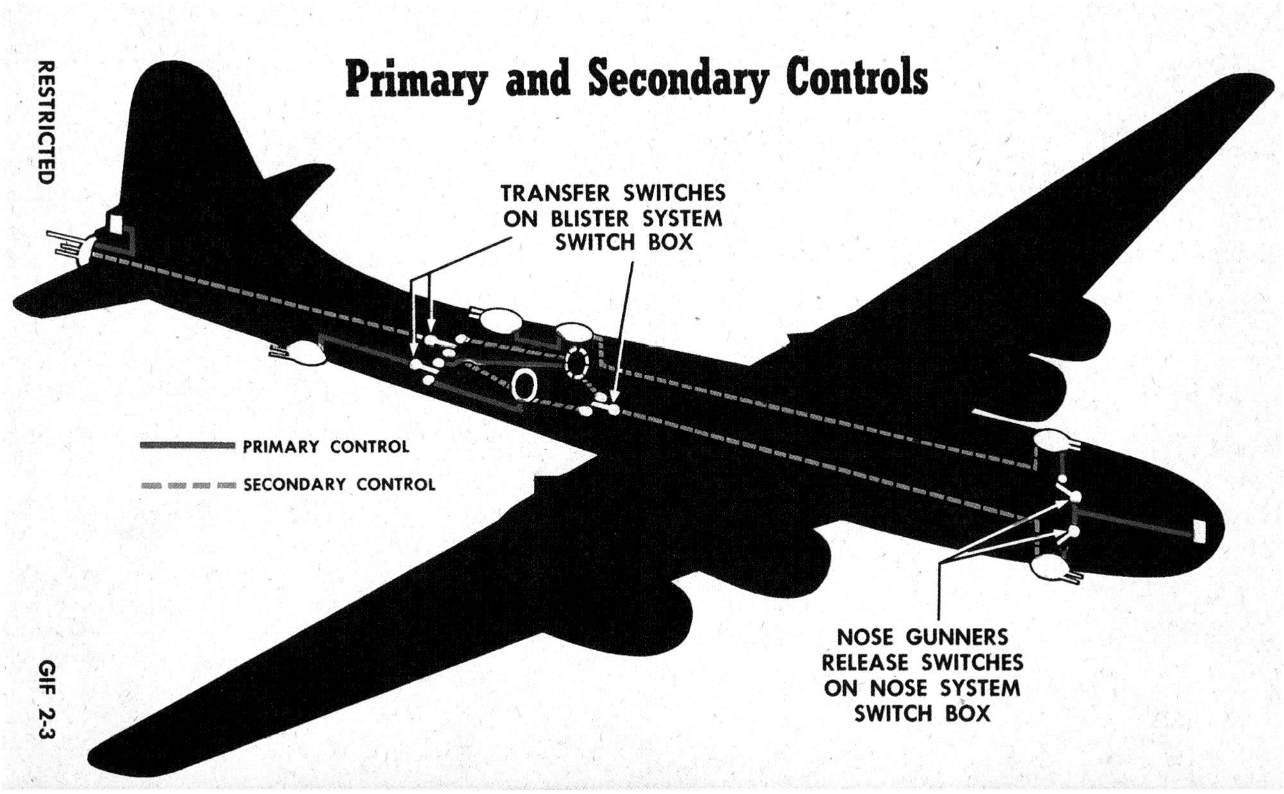
RESTRICTED
Primary and Secondary Controls
TRANSFER SWITCHES ON BLISTER SYSTEM SWITCH BOX
PRIMARY CONTROL
SECONDARY CONTROL
NOSE GUNNERS RELEASE SWITCHES ON NOSE SYSTEM SWITCH BOX
GIF 2-3

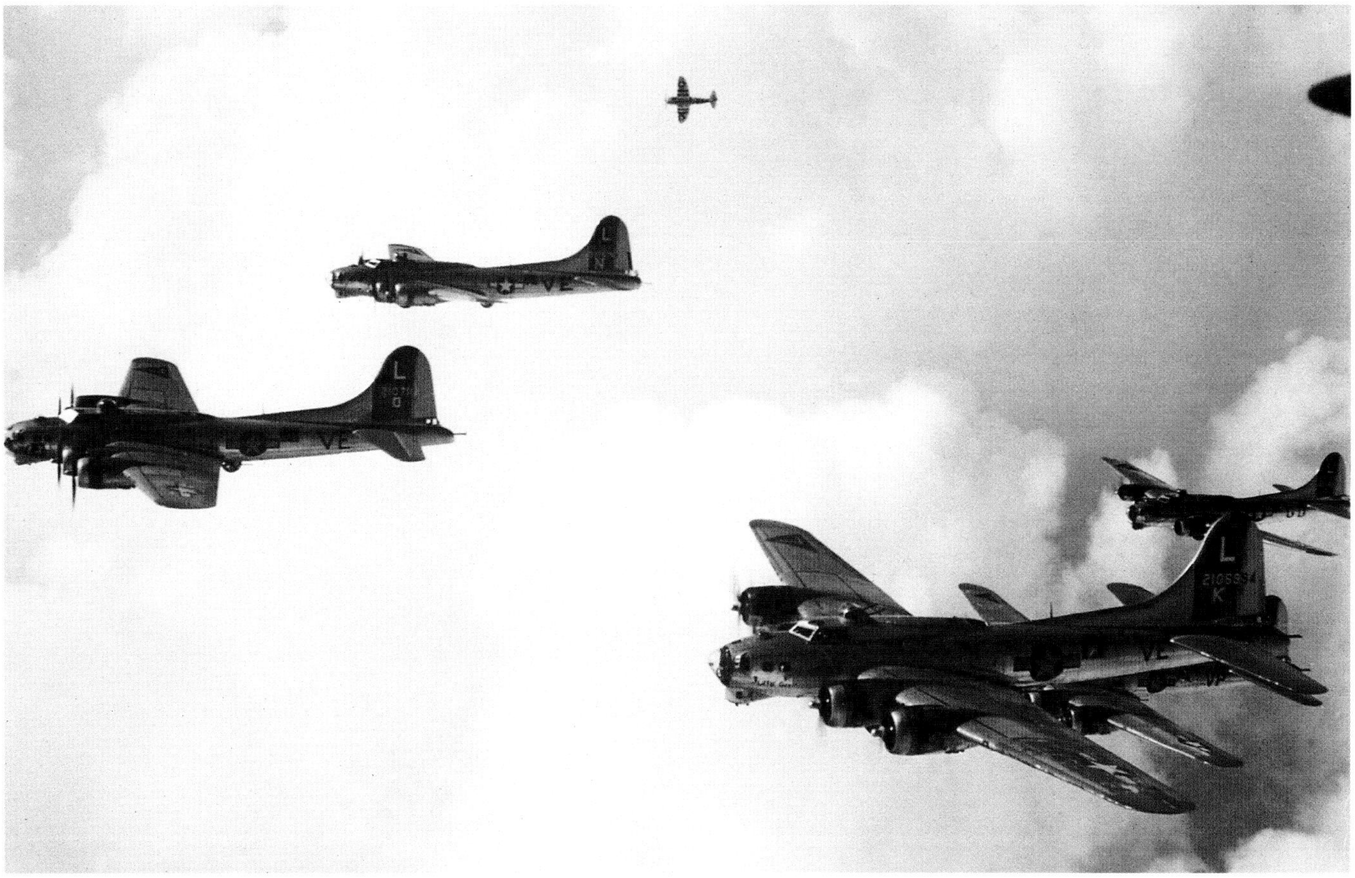

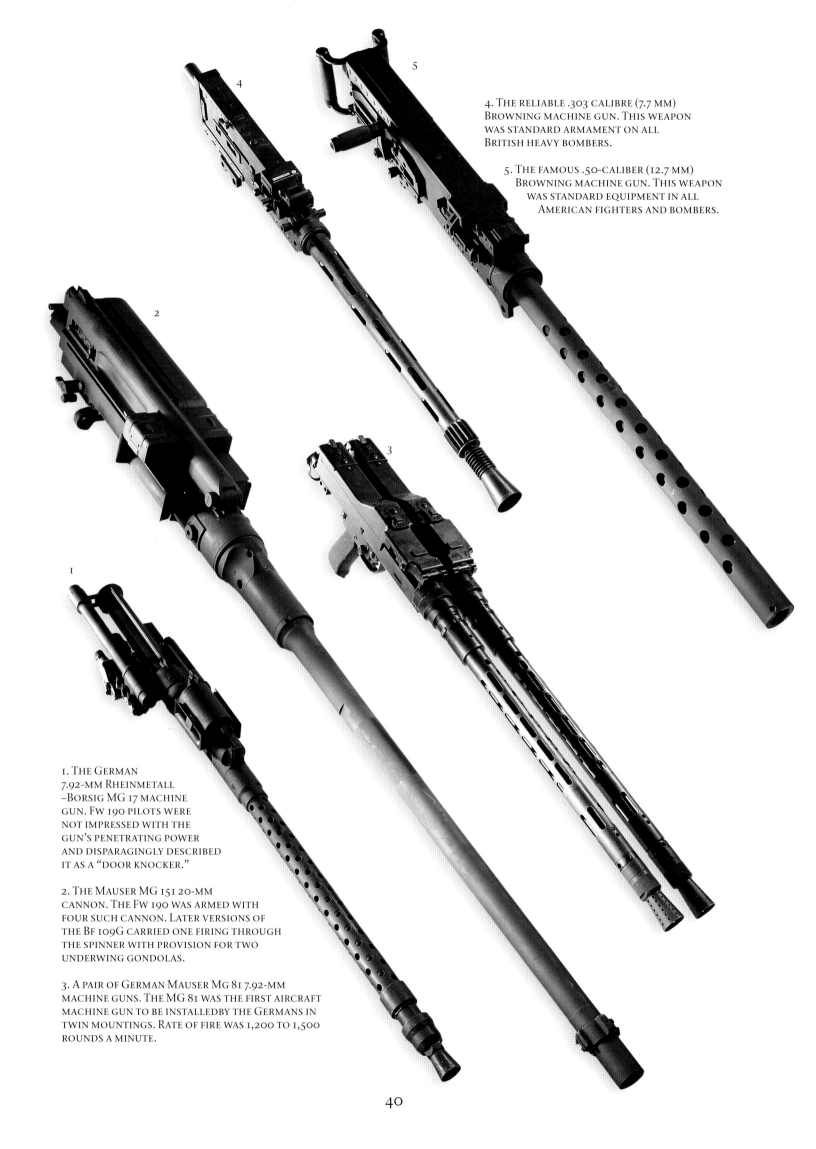

4. THE RELIABLE .303 CALIBRE (7.7 MM) BROWNING MACHINE GUN. THIS WEAPON WAS STANDARD ARMAMENT ON ALL BRITISH HEAVY BOMBERS.

5. THE FAMOUS .50-CALIBER (12.7 MM) BROWNING MACHINE GUN. THIS WEAPON WAS STANDARD EQUIPMENT IN ALL AMERICAN FIGHTERS AND BOMBERS.

1. THE GERMAN 7.92-MM RHEINMETALL –BORSIG MG 17 MACHINE GUN. FW 190 PILOTS WERE NOT IMPRESSED WITH THE GUN'S PENETRATING POWER AND DISPARAGINGLY DESCRIBED IT AS A "DOOR KNOCKER."

2. THE MAUSER MG 151 20-MM CANNON. THE FW 190 WAS ARMED WITH FOUR SUCH CANNON. LATER VERSIONS OF THE BF 109G CARRIED ONE FIRING THROUGH THE SPINNER WITH PROVISION FOR TWO UNDERWING GONDOLAS.

3. A PAIR OF GERMAN MAUSER MG 81 7.92-MM MACHINE GUNS. THE MG 81 WAS THE FIRST AIRCRAFT MACHINE GUN TO BE INSTALLEDBY THE GERMANS IN TWIN MOUNTINGS. RATE OF FIRE WAS 1,200 TO 1,500 ROUNDS A MINUTE.

Aerial Weapons

At the beginning of World War II, aircraft design had far outstripped the design of aircraft weapons. During the Battle of Britain, the famous Spitfire and Hurricane were equipped with eight .303-caliber Browning machine guns. The same .303-caliber gun that was used in the First World War. The British soon discovered that the .303-caliber machine gun, while reliable and accurate, lacked the necessary punch. After the Battle of Britain, the Spitfire's armament was increased to include two 20-mm cannon and four .303-caliber machine guns. German and Japanese fighter armament also increased. As the bomber offensive against the Axis grew, Allied bomber crews were confronted with enemy fighters armed with 20-mm and 30-mm cannon. To defend themselves, American gunners relied on the hard-hitting .50-caliber machine gun, while the British soldiered on with the reliable but light .303-caliber machine gun.

6

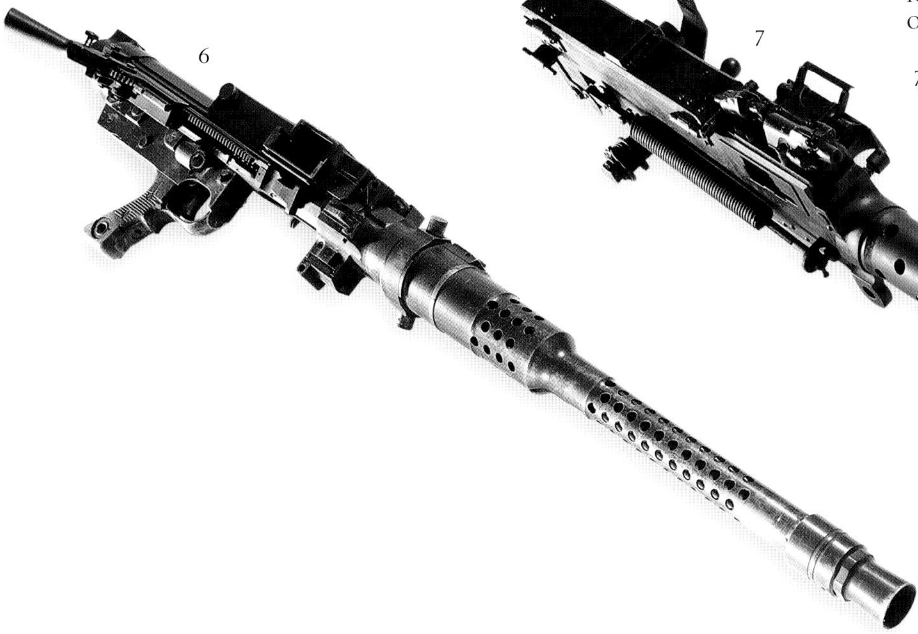

7

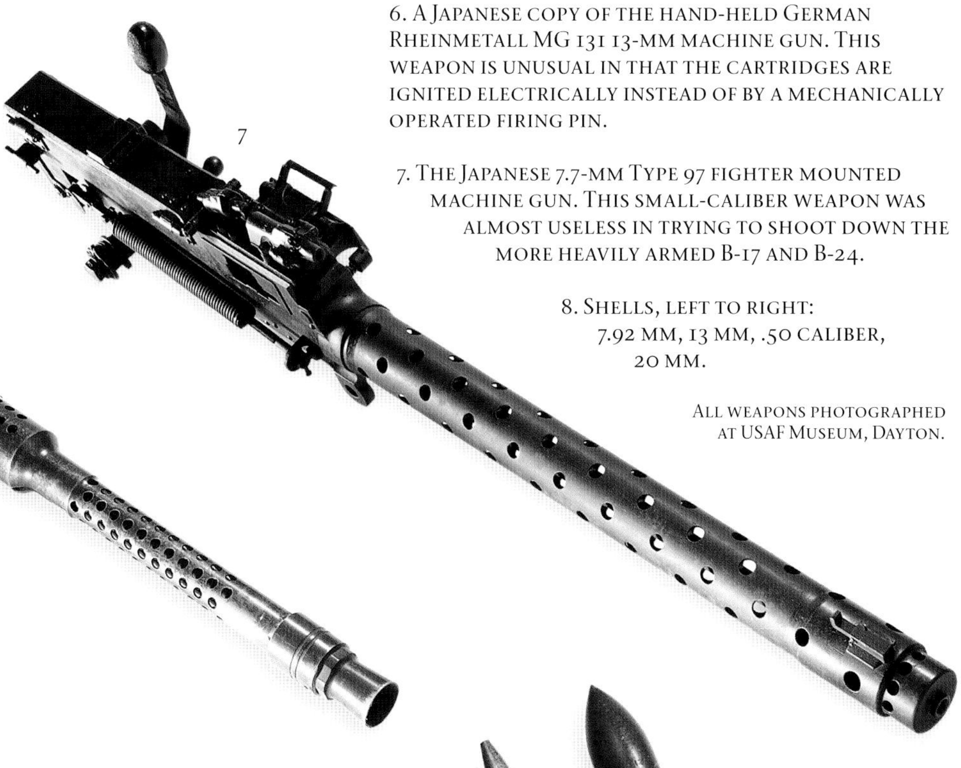

6. A Japanese copy of the hand-held German Rheinmetall MG 131 13-mm machine gun. This weapon is unusual in that the cartridges are ignited electrically instead of by a mechanically operated firing pin.

7. The Japanese 7.7-mm Type 97 fighter mounted machine gun. This small-caliber weapon was almost useless in trying to shoot down the more heavily armed B-17 and B-24.

8. Shells, left to right: 7.92 mm, 13 mm, .50 caliber, 20 mm.

All weapons photographed at USAF Museum, Dayton.

8

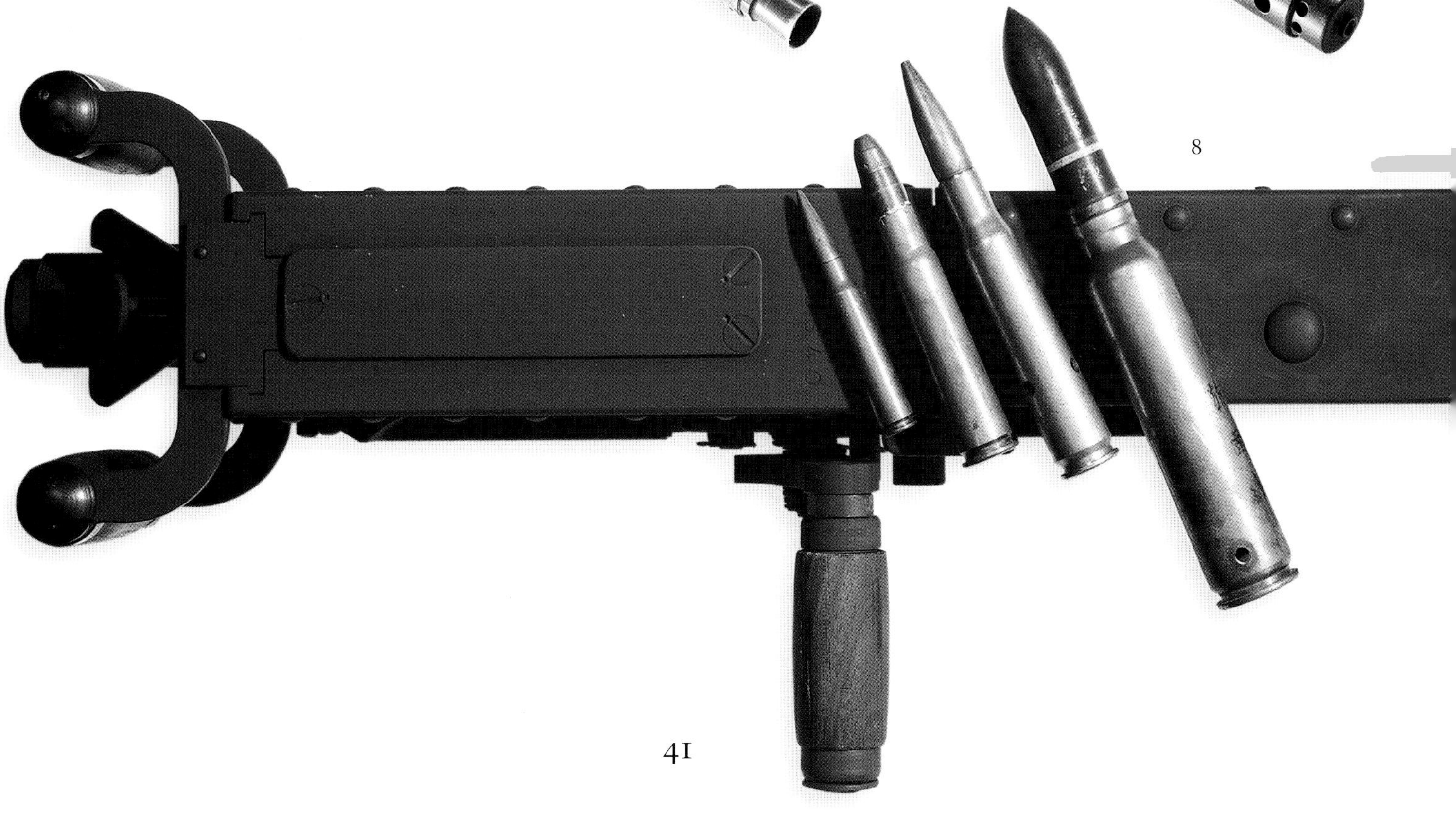

■★■ BOEING B-17 FLYING FORTRESS ■★■ CONSOLIDATED B-24 LIBERATOR

■★■ BOEING B-29 SUPERFORTRESS

FOUR ENGINE

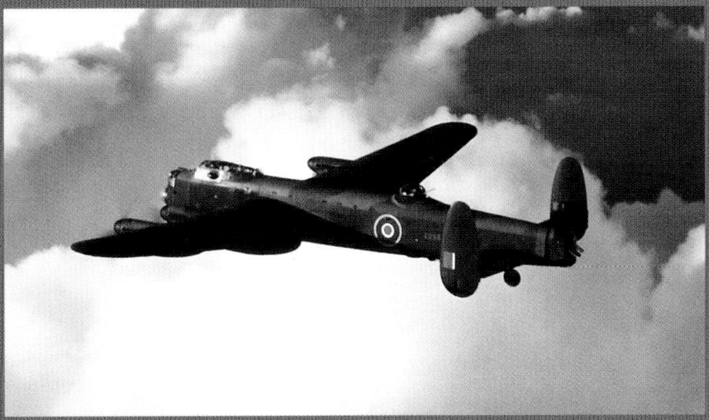

AVRO LANCASTER

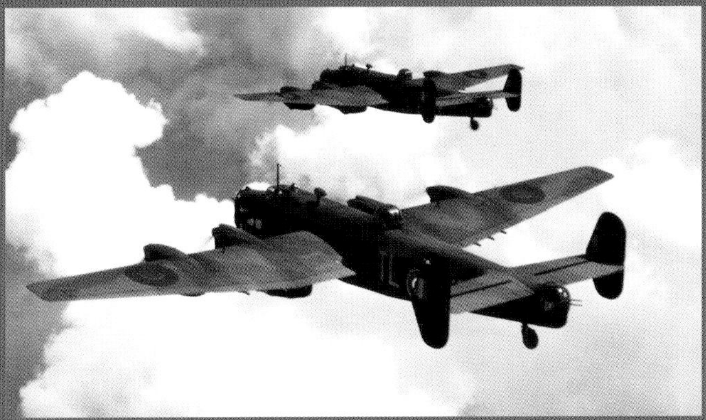

HANDLEY PAGE HALIFAX

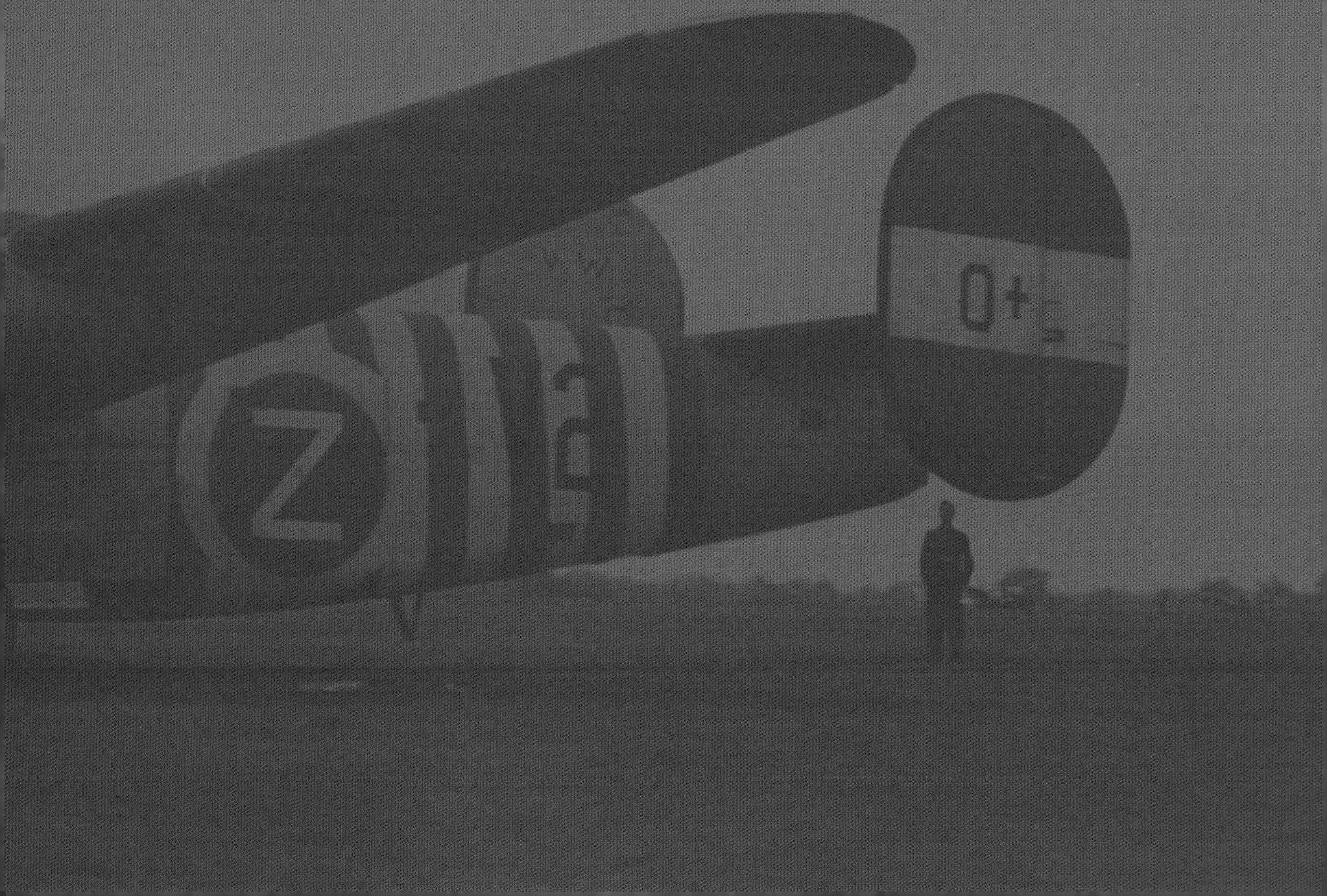

Boeing B-17
FLYING FORTRESS

THE ODDS AGAINST FORTRESS GUNNERS, ESPECIALLY THOSE WHO FLEW OVER GERMANY, WERE FORMIDABLE, BUT IN THE END THEIR BRAVERY AND THEIR LONG-RANGE ESCORTS HELPED DEFEAT THE ONCE-POWERFUL LUFTWAFFE.

No other airplane better represented the idea of the "self-defending" bomber than the B-17. In the battles high above Europe, B-17s from England and the Mediterranean played a vital role in the destruction of the Luftwaffe. Heavily armed and armored, B-17 gunners, in their many turrets and gun positions, fought their way through the world's toughest air defense system. Although many gunners did not return, the majority did, thanks to their courage and tenacity — and the rugged construction of the B-17.

The B-17 first saw combat with the Royal Air Force beginning in July 1941. Its defensive armament could not be described as fortress-like. The B-17C was armed with two hand-held .50-caliber guns in the dorsal position, two in the ventral position, one .303-caliber flexible gun in the nose and two .50-caliber guns in the two waist positions. The RAF wanted to use the B-17 as a high-altitude bomber in the hopes of avoiding enemy fighters. The results were not good. The guns had a tendency to freeze at high altitude, and losses to fighters and accidents steadily grew until only four of the original twenty B-17s remained.

The experience of the B-17C showed the Americans that the B-17's defensive armament did not live up to the Fortress name. An extensive redesign was ordered, and the first truly combat-ready B-17 was the E. The E model was equipped with a Sperry top turret, a Sperry remote lower turret (soon replaced by the Sperry ball turret), a tail turret, two guns in the waist positions and two .30-caliber guns in the nose. The next model to enter service was the B-17F. Outwardly, the only difference from the E model was the extended Plexiglas nose, but F model incorporated hundreds of changes, including new fuel tanks and more armor. The B-17F was the bomber that bore the brunt of early deep-penetration raids into Germany. Against stiffening fighter resistance, losses frequently rose above ten percent. It also became clear that the B-17s, like the early B-24s, were vulnerable to head-on attacks. The B-17G attempted to solve that problem with the addition of the Bendix chin turret, operated by the bombardier. The B-17G was now armed with thirteen .50-caliber machine guns! Even with the increased armament, the mighty B-17 was still vulnerable. In the end, it would be the introduction of long-range fighter escorts that enabled the 8th and 15th Air Forces to continue the daylight strategic bombing of Germany.

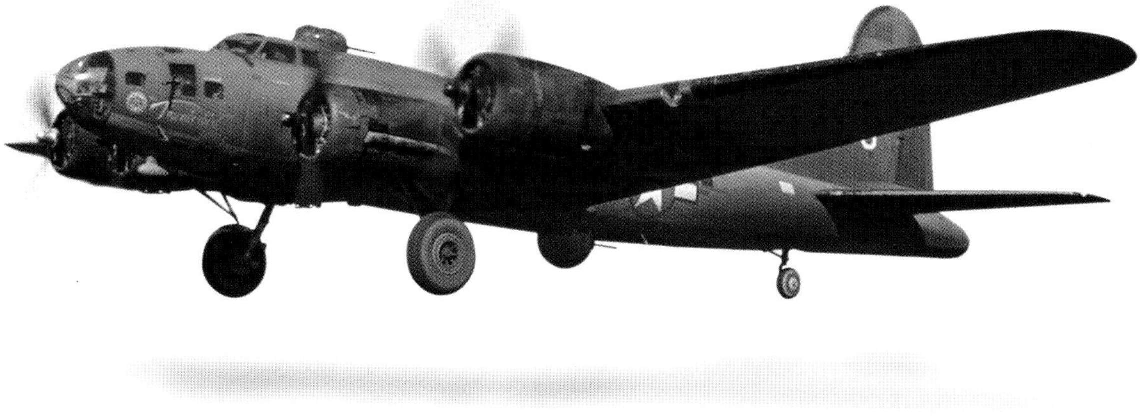

"I flew in both the E and F model. When the German fighters came in head-on, it was almost a no-miss situation. The rate of closure was so fast. The German tactics were always to get several thousand feet above the formation and get the sun behind them. Of course, it was more difficult to see them in the sun, but we were always on the lookout for them. You would look up high, to 12 and 3 o'clock, and that's where they were. Then they would dive down, and the gunners would begin calling them out. We would fire short bursts to conserve ammunition. With the head-on attack, the rate of closure was around 600 miles an hour, so you had to get your bursts in in a hurry. You could tell when we had a new gunner aboard, because he would start firing as soon a fighter would make a turn. At that point, they were still a long way away and it was just a waste of ammunition. You had to wait until the enemy fighter was about 600 or 700 yards away, and even then it was strictly a guess.

"Every fifth bullet was a tracer. That guided you. It was an excellent way of showing where you were aiming and hitting. Our tracers, experience and the firepower from all the other airplanes made us confident.

"The Germans were completely ruthless. We encountered Me 109s flying from England and in Africa. Later we began to see the Fw 190, and soon after that we encountered the yellow-nosed Fw 190s. They were the elite of the German Air Force. The first time we saw them, we were in awe. They displayed great maneuverability and daring. They attacked mostly from the front. When they came down from above, they would actually do a roll as they fired their cannons and then would disappear underneath our formation. When I saw that, I was almost spellbound and for a split second I didn't fire. It was just a beautiful sight."

1st Lieutenant Paul Blaida,
B-17 Bombardier/Gunner, 342nd Squadron,
97th Bomb Group

Top: When an aircraft was designated war weary (WW) it quickly found a role as a trainer or transport aircraft. This fully armed B-17F was used as a crew-training aircraft.
Below: Three early B-17Gs from the 390th Bomb Group form up over England and head for Germany.

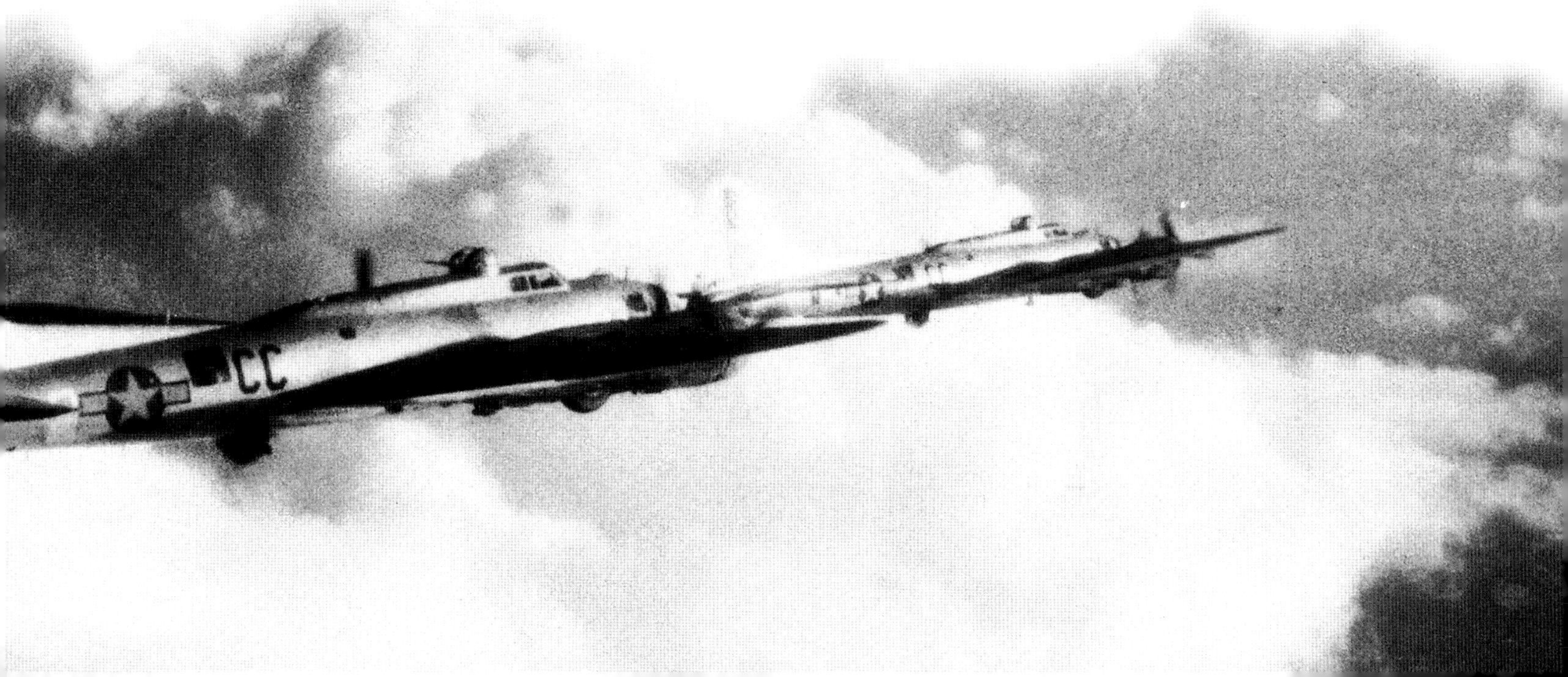
CC

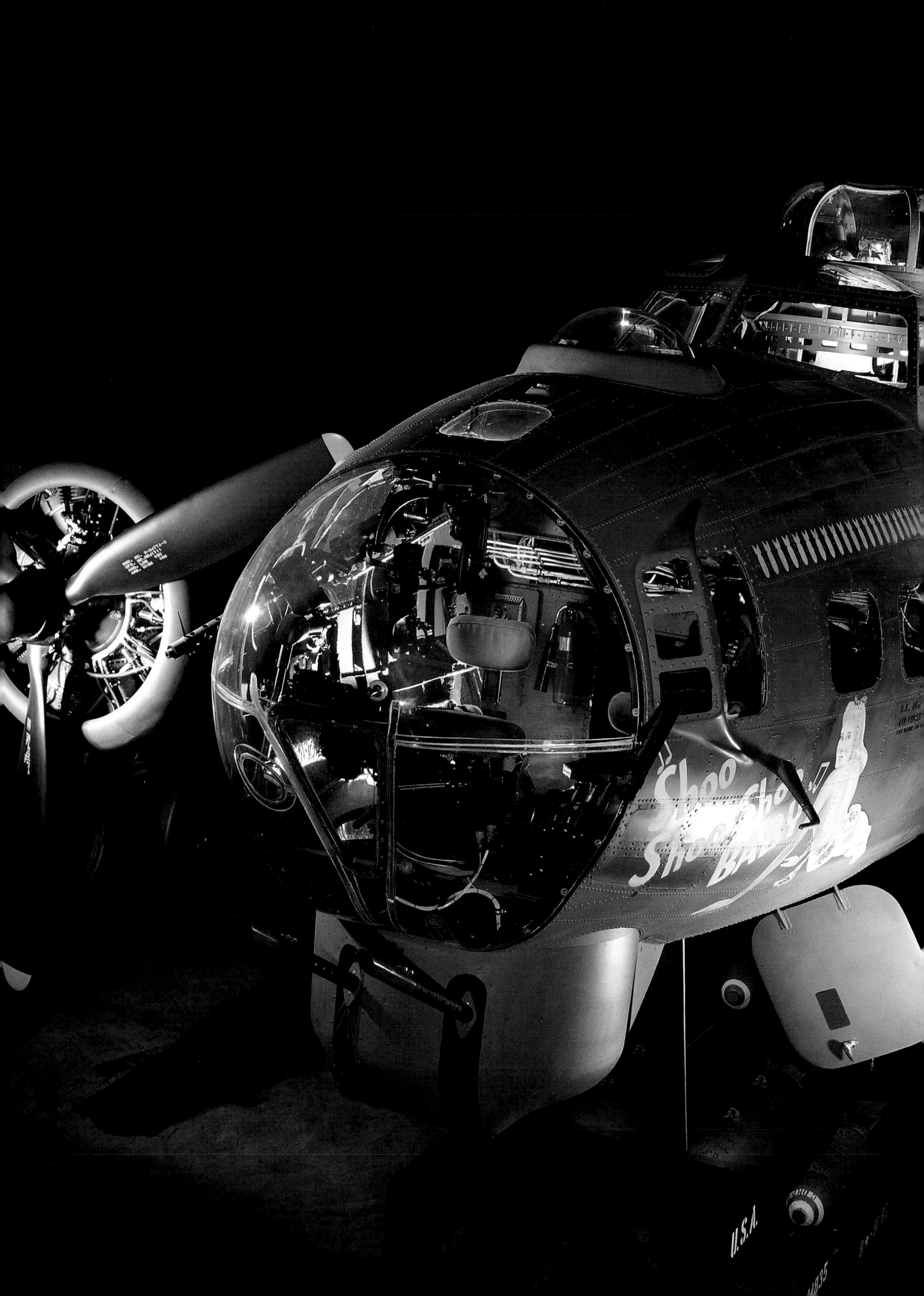
Shoo
U.S.A.

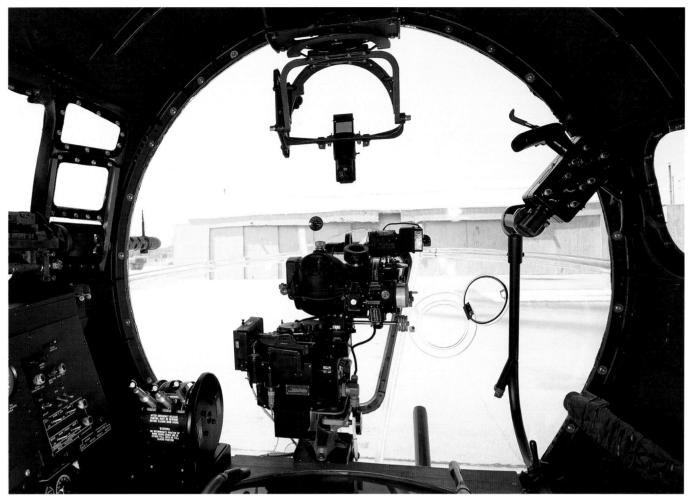

"We started wearing army helmets cut down to fit over your head. I was tall, so every time I shot straight up I had to squat down because the gunsight would press against my helmet. I cut more sections out, but in the end, I didn't wear it because, basically, there was hardly anything left.

"On this one mission, a German fighter strafed over top of us with 20-mm shells. He hit the top of the aircraft and left a row of holes. After that, every time I moved the turret past a certain point, the radicals in the sight would jump up and down and then smooth out. We were still under attack, so I had to keep the turret rotating. Every time I went back across, it would cause me problems, but it wasn't as bad. Finally, the fighter attacks ended and I didn't have to move the turret. When I got back to the airbase, I told the mechanics to check the turret, that I thought it had been hit by a shell. That night they came by, woke me, and showed me what they had found. There was a 20-mm shell lodged in the ring gear. Every time I rotated the turret I was

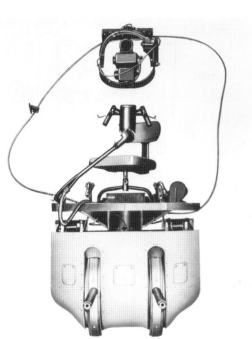

LEFT: THE INTRODUCTION OF THE BENDIX CHIN TURRET ON THE B-17G WAS A GREAT EXAMPLE OF AMERICAN DESIGN AND PRODUCTION INGENUITY. IN JUST A FEW SHORT MONTHS, BENDIX HAD PRODUCED A TWIN-GUN FRONT TURRET THAT RAISED MORALE AND GAVE GERMAN FIGHTER PILOTS A NASTY SHOCK.
USAF MUSEUM, DAYTON.

ABOVE RIGHT: THE BENDIX ELECTRICALLY POWERED REMOTE-CONTROLLED POWERED TURRET.

TOP: THE BOMBARDIER MANNED THE CHIN-MOUNTED BENDIX REMOTE-POWERED TURRET WHILE THE NAVIGATOR TOOK CARE OF THE TWO MANUALLY OPERATED CHEEK GUNS (THE LEFT GUN IS VISIBLE). THE HANDLEBAR CONTROLLER TOP (LEFT) WAS PULLED INTO PLACE DIRECTLY IN FRONT OF THE BOMBARDIER WHEN NEEDED. THE BOMBARDIER THEN TRACKED HIS TARGETS WITH THE GUNSIGHT (N-8 OR N-6A).

grinding it down. Had I continued rotating the turret I would have ground down to the primer and it would probably have exploded. They said they were going to get it out of the turret and give it to me, and I said, oh no, you guys just keep that shell, I don't want any part of it.

"I remember one mission where the German fighters were coming in from around 10 and 11 o'clock. They shot down our lead plane, and I was shooting with all I could give. I got my sights on this one fighter and gave him a good burst. I think I saw him smoking, but there were two or three other fighters right behind him. As soon as one went by, you had to start looking for the others. You never had time to see if you had shot someone down.

"You had to be real calm and not get overexcited when you were shooting. If you got real excited and didn't know what you were shooting at, you were a menace. There were several occasions where other bombers were hit by gunners. It was mostly waist gunners. If one plane slowed up and broke formation, there was a real danger of it being hit by other gunners. Every gunner has a patch of sky in which to work with, but if a B-17 slips into your patch of sky and you're concentrating on incoming fighters, there's a real danger of hitting that aircraft."

DON ATKINSON,
B-17 TOP TURRET GUNNER
418TH SQUADRON, 100TH BOMB GROUP

TOP: THE TOP TURRET GUNNER'S VIEW. EARLY MODELS OF THE SPERRY TOP TURRET WERE HAMPERED BY HEAVY METAL FRAMING RUNNING THROUGH THE PERSPEX. LATER MODELS WERE EQUIPPED WITH A CLEAR PLEXIGLAS DOME.

BELOW: THE CHEEK GUNS MOUNTED IN THE B-17 WERE OF LIMITED USE AND NOT VERY EFFECTIVE. THEIR FIELD OF FIRE WAS RESTRICTED TO A VERY SMALL PATCH OF SKY. THIS .50-CALIBER MACHINE GUN USES A BALL JOINT MOUNTING IN A B-17F.

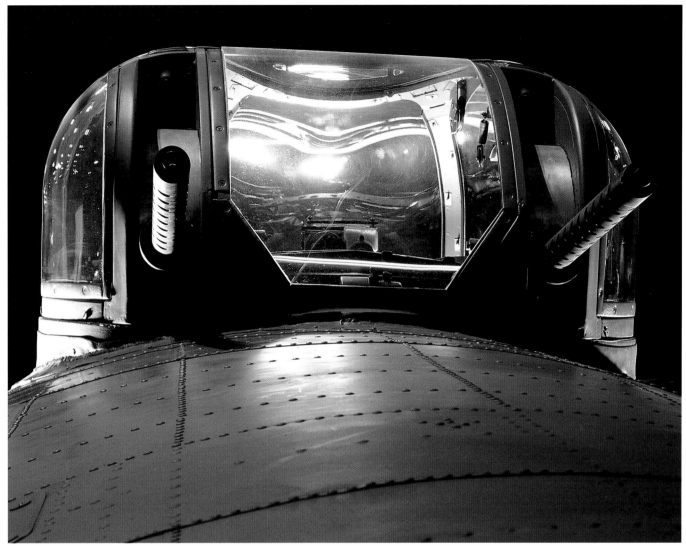

THE SPERRY TOP TURRET WAS VERY EFFICIENT AND, WITH AN EXPERIENCED GUNNER, WAS A DEADLY THREAT TO ANY ATTACKING FIGHTER. THIS TURRET CARRIED 375 ROUNDS OF AMMUNITION AND COULD BE RELOADED IN FLIGHT.
USAF MUSEUM, DAYTON.

"The thing about the waist position in the B-17 was, there was another gunner there with you and you could say it was kind of crowded. Later on in the war, they took one of the gunners out. But when I flew, the waist guns were staggered; one of them was forward, the other was aft. It gave you more room to maneuver. It wasn't really that bad.

"The .50-caliber had an air-cooled jacket wrapped around the barrel. You were only supposed to fire the gun in short bursts. If you bore down and shot a long burst, there was a real good chance of warping the barrel. That wasn't good. The secret was to try and shoot short bursts and try to hit 'em with that.

"The waist position, by the time I got over there, was not really a great position in terms of getting in some good shots. You were flying in such a tight formation that, when a German fighter got in between you and another B-17, there was chance of hitting one of your own aircraft.

"When I first got over there, the German fighters were coming in from the rear. When they came in from the rear, the only way they could hit you was to point their guns directly at you. After they made their pass, they would roll onto their bellies and dive. That's when you had good chance to shoot 'em down. It wasn't long after that they started coming in from the front. When that happened, they went by you so fast you didn't have a chance. The only time the waist gunner really got a chance to do a whole lot of shooting was when you became a straggler. That's when the fighters would come in from every direction and every gun in the ship became active.

"All the missions scared me to death. Whether you had fighters or not, you still had to fly through the flak. Flak was what really got you thinking, but I found a way to suck it up and go."

WILLIAM J. HOWARD,
B-17 WAIST GUNNER,
351ST SQUADRON, 100TH BOMB GROUP

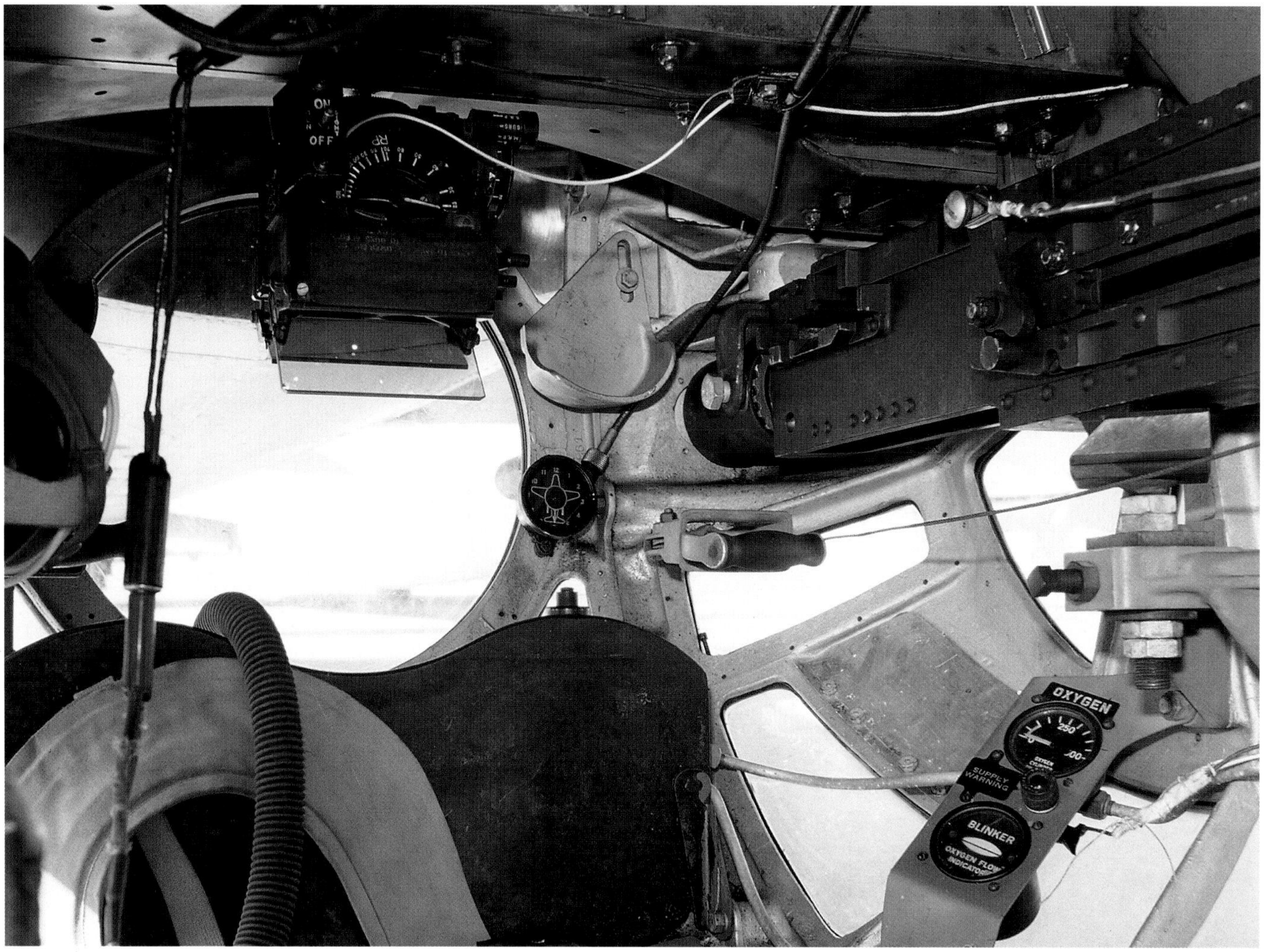
ON
OFF
OXYGEN
SUPPLY WARNING
BLINKER
OXYGEN FLOW INDICATOR

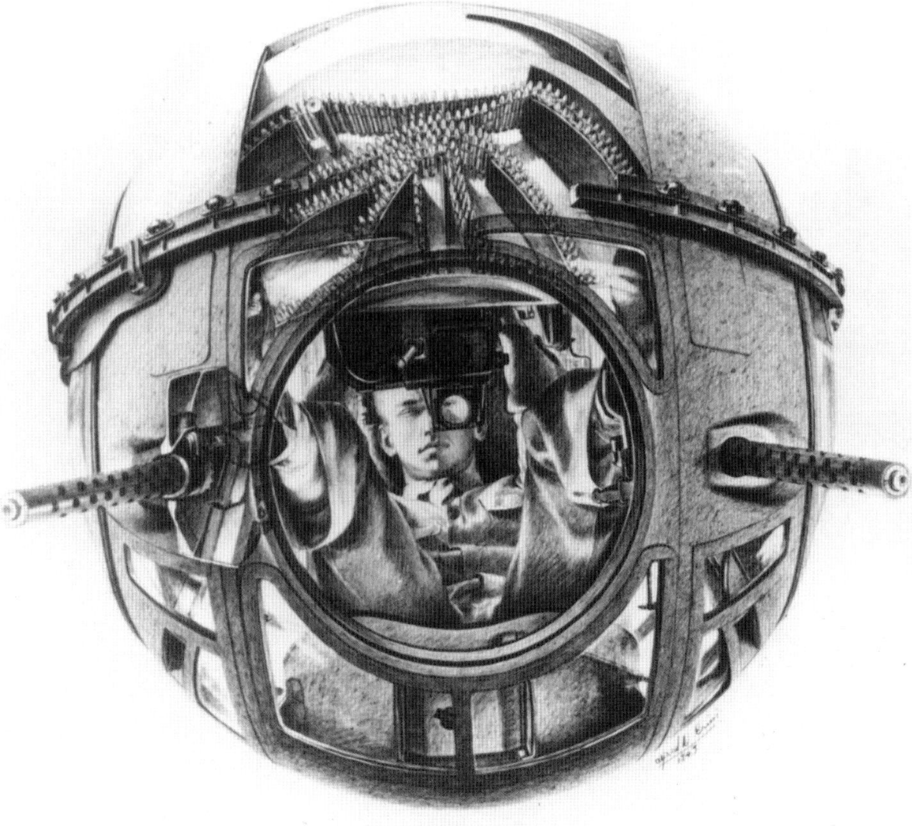

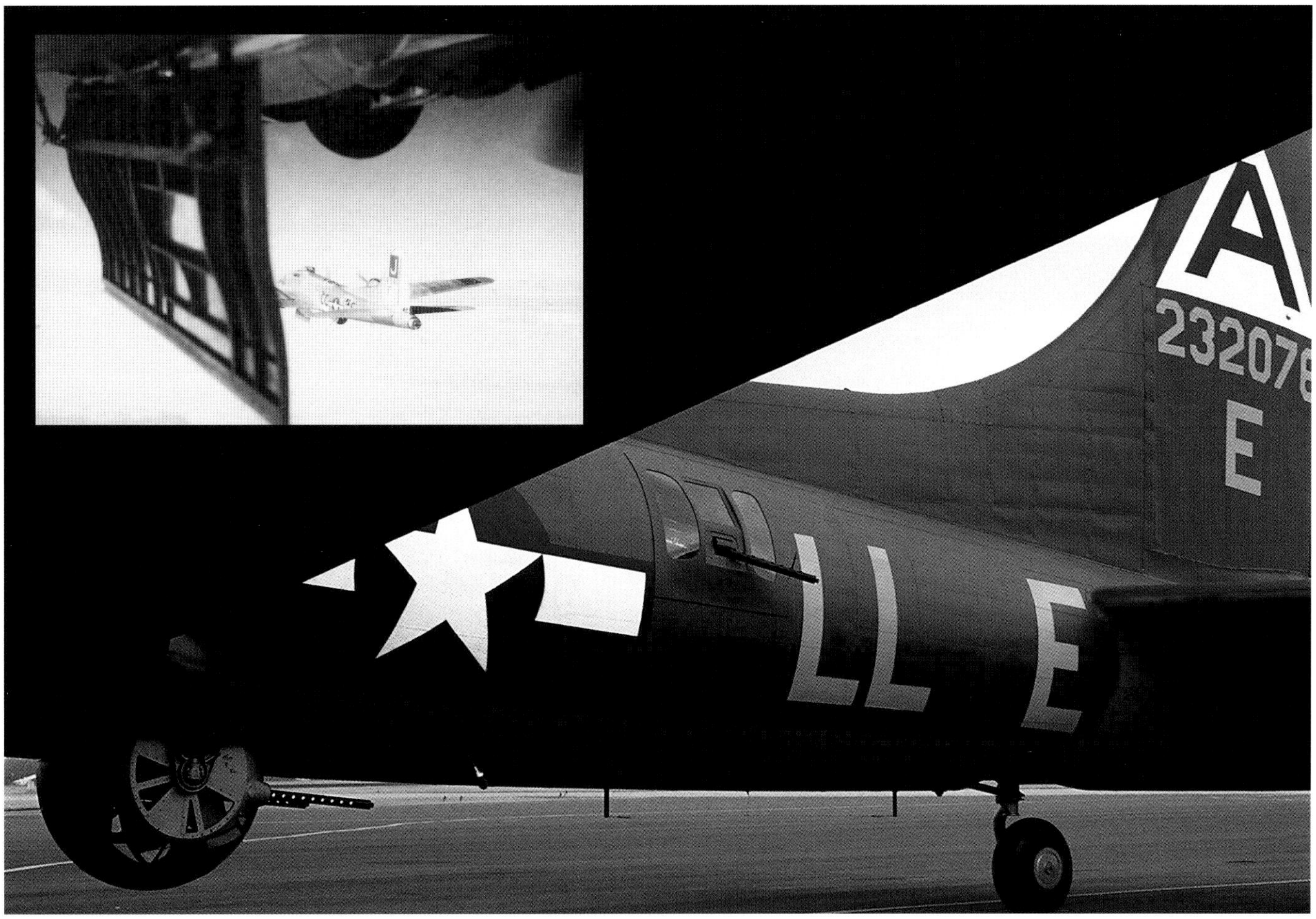
A
232076
E
LL E

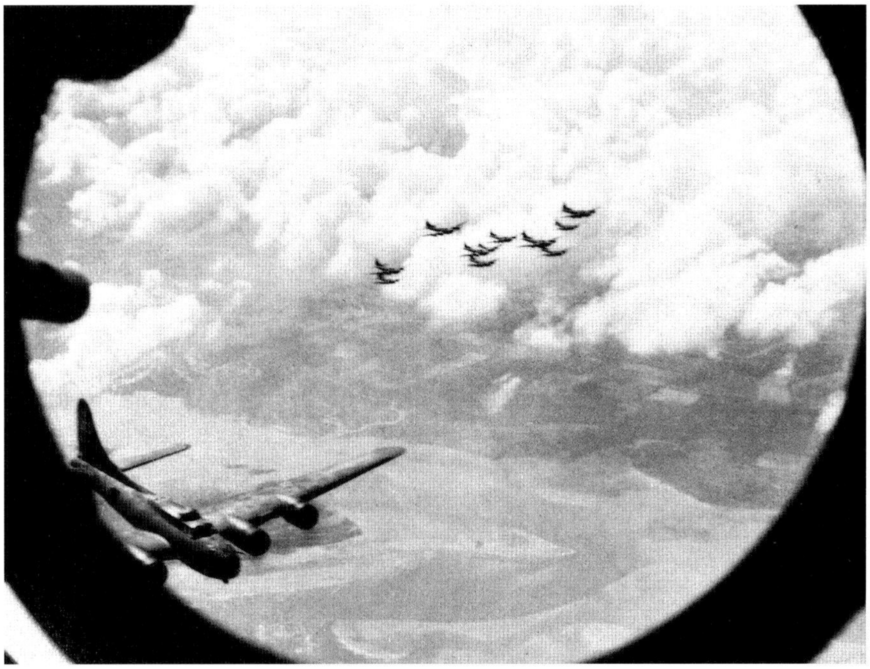

Top left: The extrememly cramped interior of the Sperry ball turret.
Both the B-17 and the B-24 were equipped with this type of turret. Exposed and vulnerable as it might have seemed, the ball turret was in fact the safest position in the B-17 and B-24.

Bottom left: Sperry ball turret. This wartime drawing shows the ammunition storage system.

Top: B-17G, *Shoo Shoo Baby*, combat veteran of twenty-two missions.
USAF Museum, Dayton.

Inset: View from the ball turret. With the bomb-bay doors open, B-17s of the 390th Bomb Group begin their bomb run.

Bottom: The ball gunner's view. 390th Bomb Group B-17s on their way to Germany.

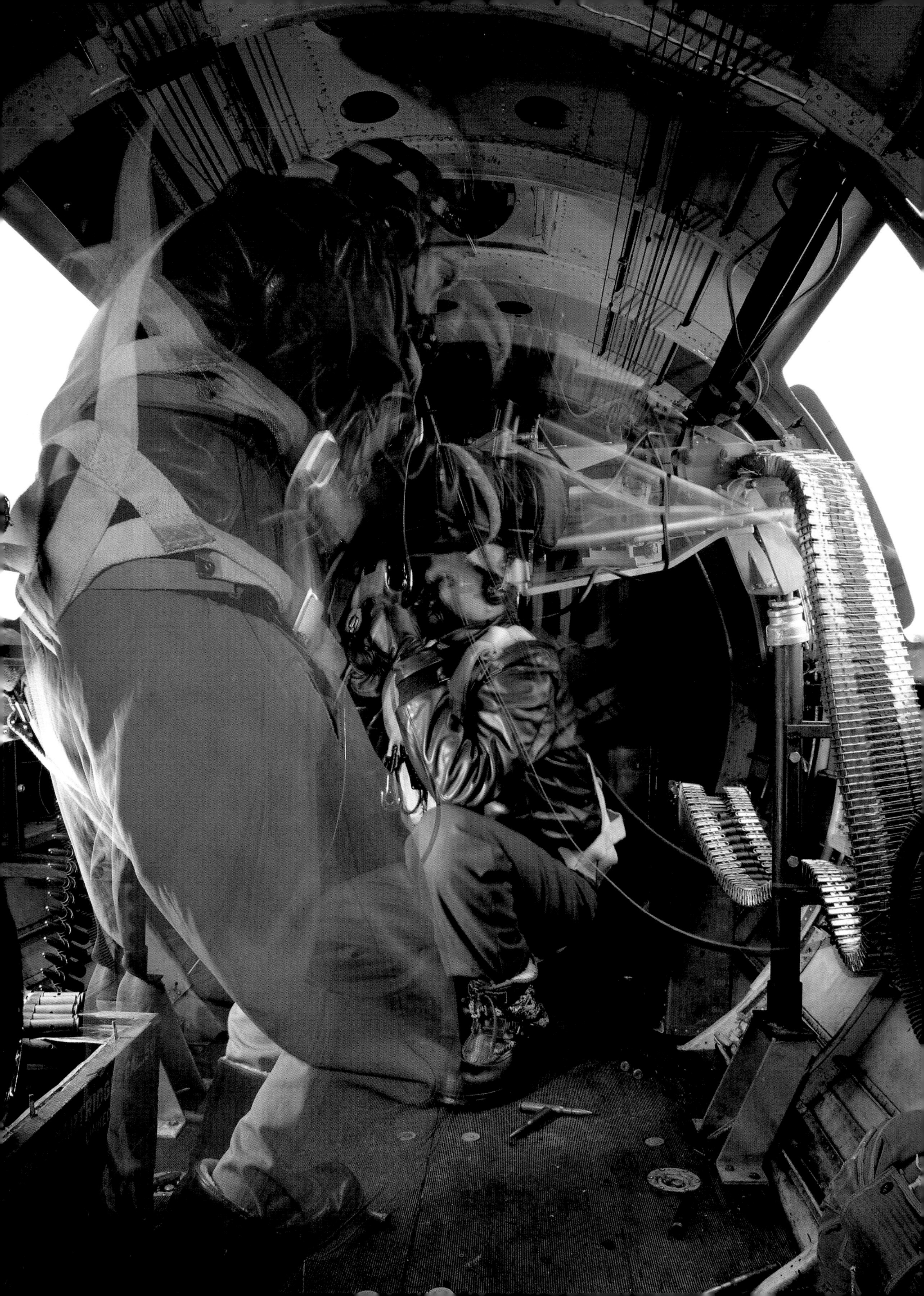

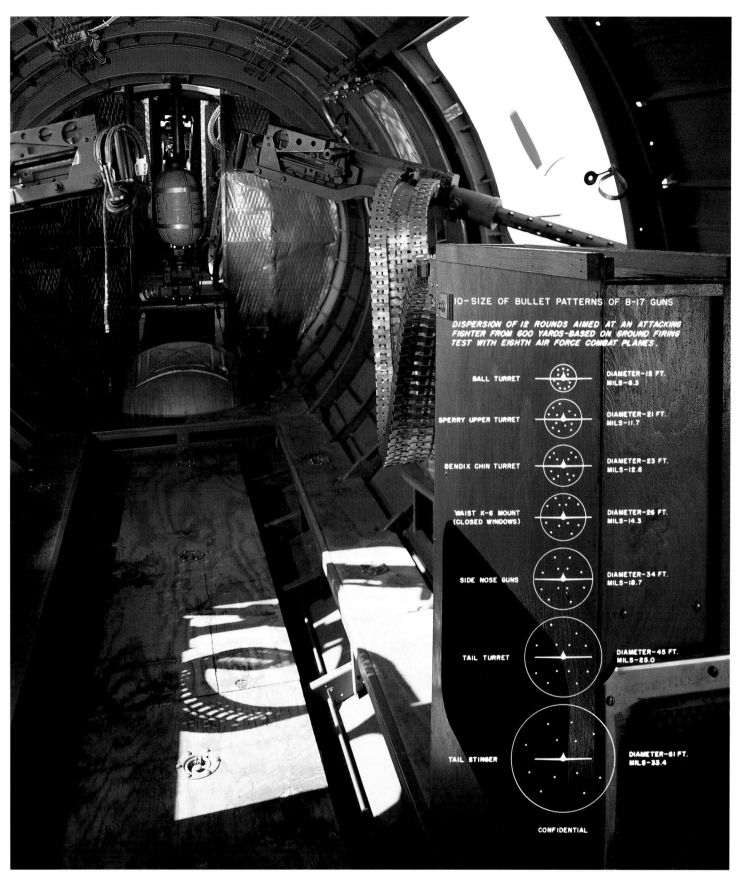

Within the photograph, the inset panel reads:

IO-SIZE OF BULLET PATTERNS OF B-17 GUNS

DISPERSION OF 12 ROUNDS AIMED AT AN ATTACKING
FIGHTER FROM 600 YARDS-BASED ON GROUND FIRING
TEST WITH EIGHTH AIR FORCE COMBAT PLANES.

BALL TURRET	DIAMETER-15 FT. MILS-8.3
SPERRY UPPER TURRET	DIAMETER-21 FT. MILS-11.7
BENDIX CHIN TURRET	DIAMETER-23 FT. MILS-12.6
WAIST K-6 MOUNT (CLOSED WINDOWS)	DIAMETER-26 FT. MILS-14.3
SIDE NOSE GUNS	DIAMETER-34 FT. MILS-18.7
TAIL TURRET	DIAMETER-45 FT. MILS-25.0
TAIL STINGER	DIAMETER-61 FT. MILS-33.4

CONFIDENTIAL

Left: Time-lapse photography shows just how little room there was in the B-17 waist gun position.
Hitting an enemy fighter from the waist position was extremely difficult.
A quick burst at a passing enemy fighter was the most a gunner could hope for.

Above: The B-17 waist gunner's position. Exposed to the elements, the waist gunner had to battle frostbite
as well as the enemy. The large plywood box contained between 500 and 750 rounds of ammunition.

Inset: Size of bullet patterns of B-17 guns.

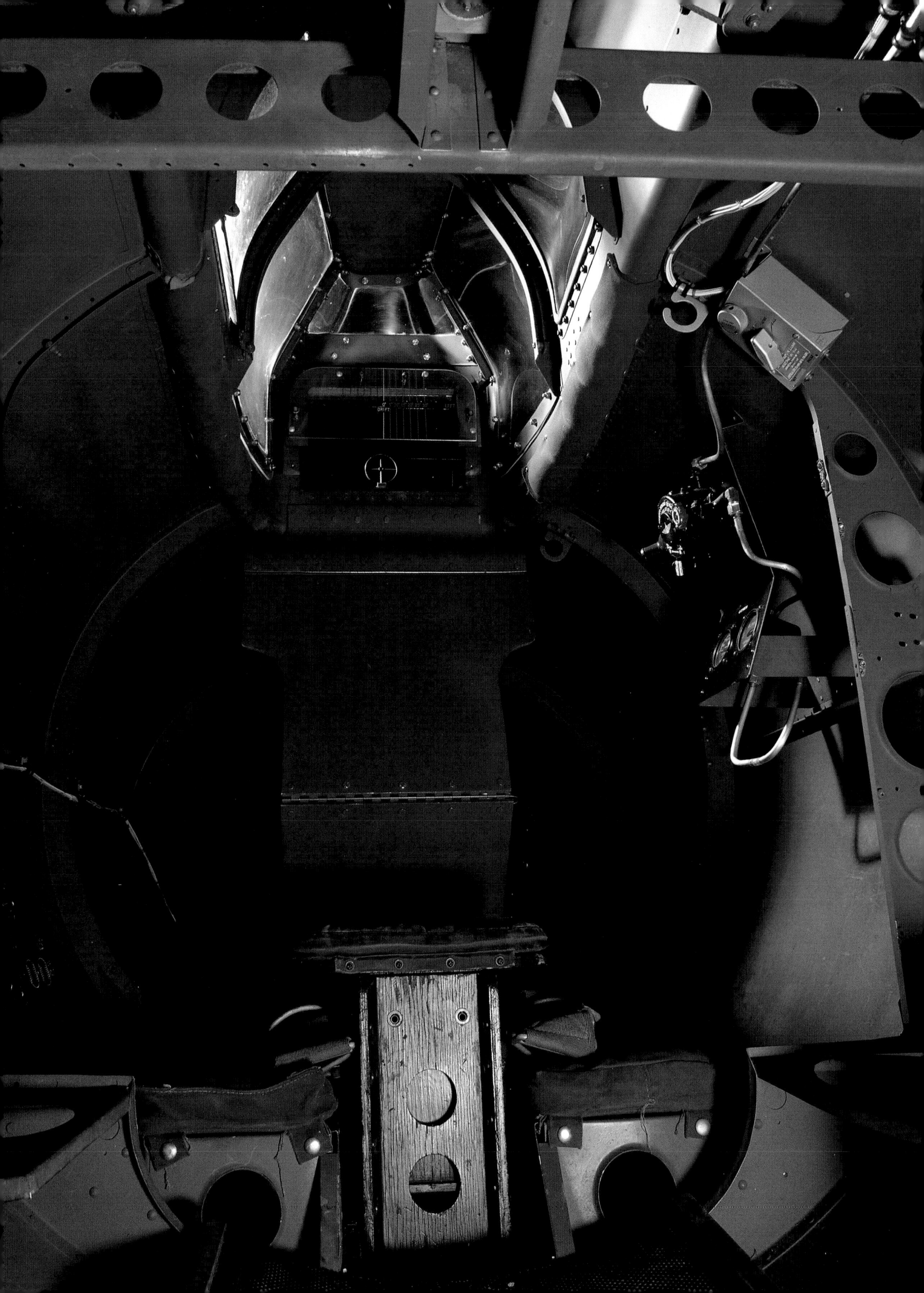
5
DRIFT

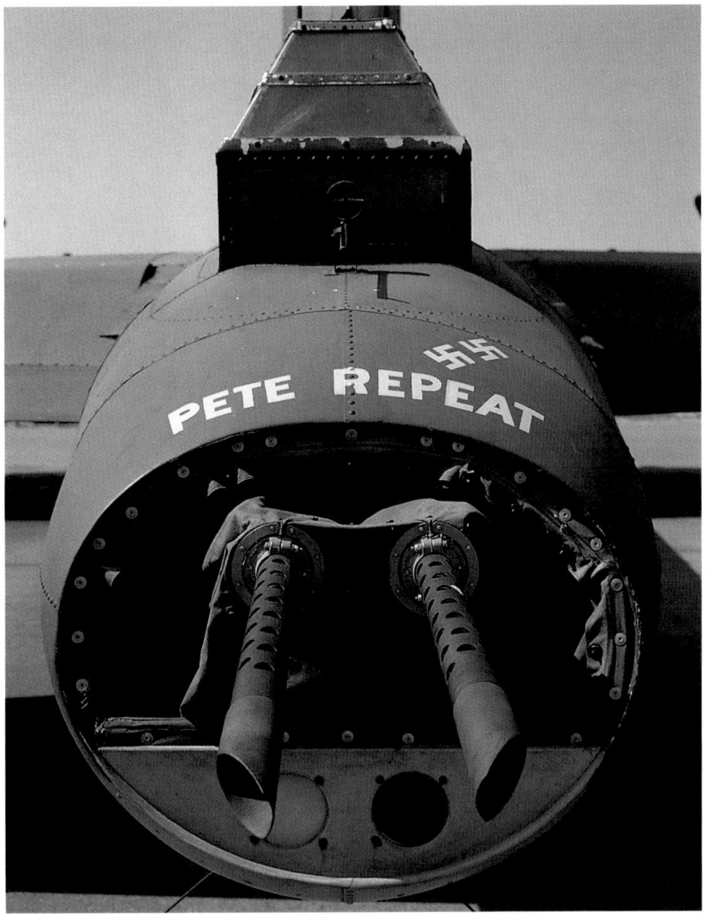
PETE REPEAT

LEFT: LOCATED BEHIND THE RUDDER, THE TAIL DEFENSE POSITION WAS EXTREMELY CRAMPED BUT VITALLY IMPORTANT. HERE THE GUNNER WAS PROVIDED WITH TWIN .50-CALIBER BROWNING MACHINE GUNS ON A HAND-OPERATED SWIVEL MOUNTING. EACH GUN WAS EQUIPPED WITH 200 ROUNDS OF AMMUNITION.
USAF MUSEUM, DAYTON.

ABOVE: THE B-17'S STINGER TAIL.

"I was an armorer/gunner flying in the tail position of the B-17. The position was rather cramped and very cold, with temperatures of anywhere from twenty to forty degrees below zero. With your oxygen mask, flak suit, steel helmet and parachute, you were not very comfortable. You didn't really sit in the tail turret, you actually had to kneel. There was a bicycle seat, and on each side of the seat there were pads where you placed your knees.

"On this one memorable mission we had good support from our 'little friends,' who engaged the Germans in a dogfight, to which I had a ringside seat. We were flying in the tail-end-Charlie position of the formation when the next thing I noticed was an Me 109 level at about 7, 8 o'clock and in a position where I couldn't reach him with my guns. He lowered his gear and flaps and flew along with us. He sat there while the ball turret gunner and I instructed the pilot to move the plane so we could get a clear shot, but the German pilot knew the blind spot of the B-17 very well. I watched as he took off his oxygen mask and smoked a cigarette. After about two or three minutes, he pulled up his gear and dropped below like a rock. He apparently had run out of ammunition. That was the closest I ever got to seeing the face of the enemy."

S/SGT. EARL R. WILBUR, B-17 TAIL GUNNER,
350TH SQUADRON, 100TH BOMB GROUP

"There was one mission where we went across the North Sea and hit cloud cover. The formation scattered, and I believe a Me 410 got onto our tail. Of course, I called it off. He was in a pursuit curve. If he keeps his nose on you and you're going forward, he's got to follow that curve. When he came in on us, I opened fire along with our top turret. He backed off. I don't know if we hit him or not, but before he left, he let two rockets go and then peeled off."

ROBERT STACHEL, B-17 TAIL GUNNER,
351ST SQUADRON,
100TH BOMB GROUP

Consolidated B-24
LIBERATOR

THE LIBERATOR WAS THE ONLY SECOND WORLD WAR BOMBER TO BE EQUIPPED WITH FOUR
GUN TURRETS EACH PRODUCED BY A DIFFERENT COMPANY.

The Liberator first saw service with the Royal Air Force beginning in March 1941. The first version, designated LB-30, was used in the transport role with the newly instituted Trans-Atlantic Return Ferry Service. The first turret-armed Liberator was the Mk II. Its armament consisted of eleven .303-caliber machine guns — eight of the guns were concentrated in two turrets. The B-24 Mk II was equipped the Boulton Paul Type A and Type E turrets. The Type E was a tail turret with four .303-caliber machine guns. This turret also equipped the Halifax bomber. The Type A top turret was also a four-gun design from Boulton Paul and was used in the Boulton Paul Defiant, Halifax, Albemarle, and

late-mark Ventura and Baltimore aircraft.

The first American B-24 to use domestic turrets was the B-24C model. It was armed with a Consolidated tail turret and a Martin upper turret located just aft of the cockpit. These turrets carried two .50-caliber machine guns. There were also three hand-held .50-caliber machine guns — one in the nose and two in the waist position.

Immediately following the B-24C was the D model, which added two more .50-caliber guns in the nose and later, the Sperry ball turret. The B-24D would serve in every theater, and in 1942–43, it was the most important long-range bomber in the Pacific. It was also used in the anti-submarine role by the British, and by late 1942, fifteen

THIS B-24J'S BOMBING DAYS ARE OVER. THE COLORFUL PAINT JOB IDENTIFIES IT AS A FORMATION
ASSEMBLY SHIP OF THE 491ST BOMB GROUP. THIS MODEL IS ARMED WITH AN EMERSON NOSE TURRET,
MARTIN TOP TURRET AND CONSOLIDATED TAIL TURRET. THE SPERRY BALL TURRET HAS BEEN REMOVED.

American anti-submarine squadrons were equipped with the type. It was the D model that first saw service with the 8th Air Force in Europe.

Early combat reports showed that the B-24, like the B-17, was vulnerable to head-on attack. The three hand-held guns, manned by the navigator/gunner and bombardier/gunner, could not cope. After the D model came the J model, which was equipped with the Emerson Model 111 front turret and 127 tail turret. It was also the only American bomber ever equipped with turrets made from five different turret manufacturers: Consolidated, Emerson, Martin, Sperry, and Erco.

“I manned the front turret in a B-24. First one over the target. That’s one thing about it. In the nose turret, you’re flying. You’re up in the air by yourself.

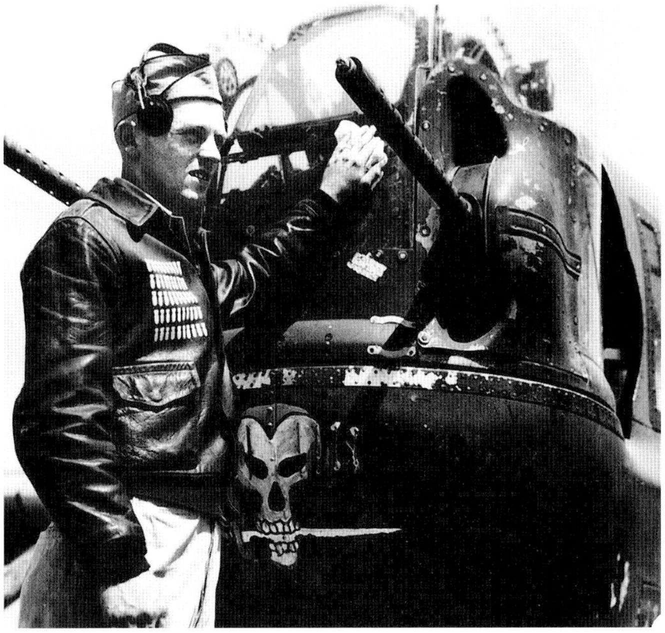

A young gunner polishes the perspex on his Emerson 127 front turret. The electrically powered Emerson was equipped with two .50-caliber machine guns with 500 rounds per gun.

“From India we would fly to Burma and dive-bomb bridges over the Burma/Siam railroad in our B-24. When we got back to England and talked to the fighter pilots, they’d say, ‘Bullshit, you’re not going to tell me you can dive-bomb in a B-24.’ But it was very true. We would start our dive from 1,500 feet, drop down to 500 feet and then drop our bombs over the bridge. Of course, the Japanese would be shooting at you and you would see the muzzle flash from their machine guns. You make a note of that and, on the next run, you would give them a good squirt. Most of the time we would make two, even three runs on a bridge.”

Rene St. Pierre,
B-24 Front Turret
Gunner, RAF

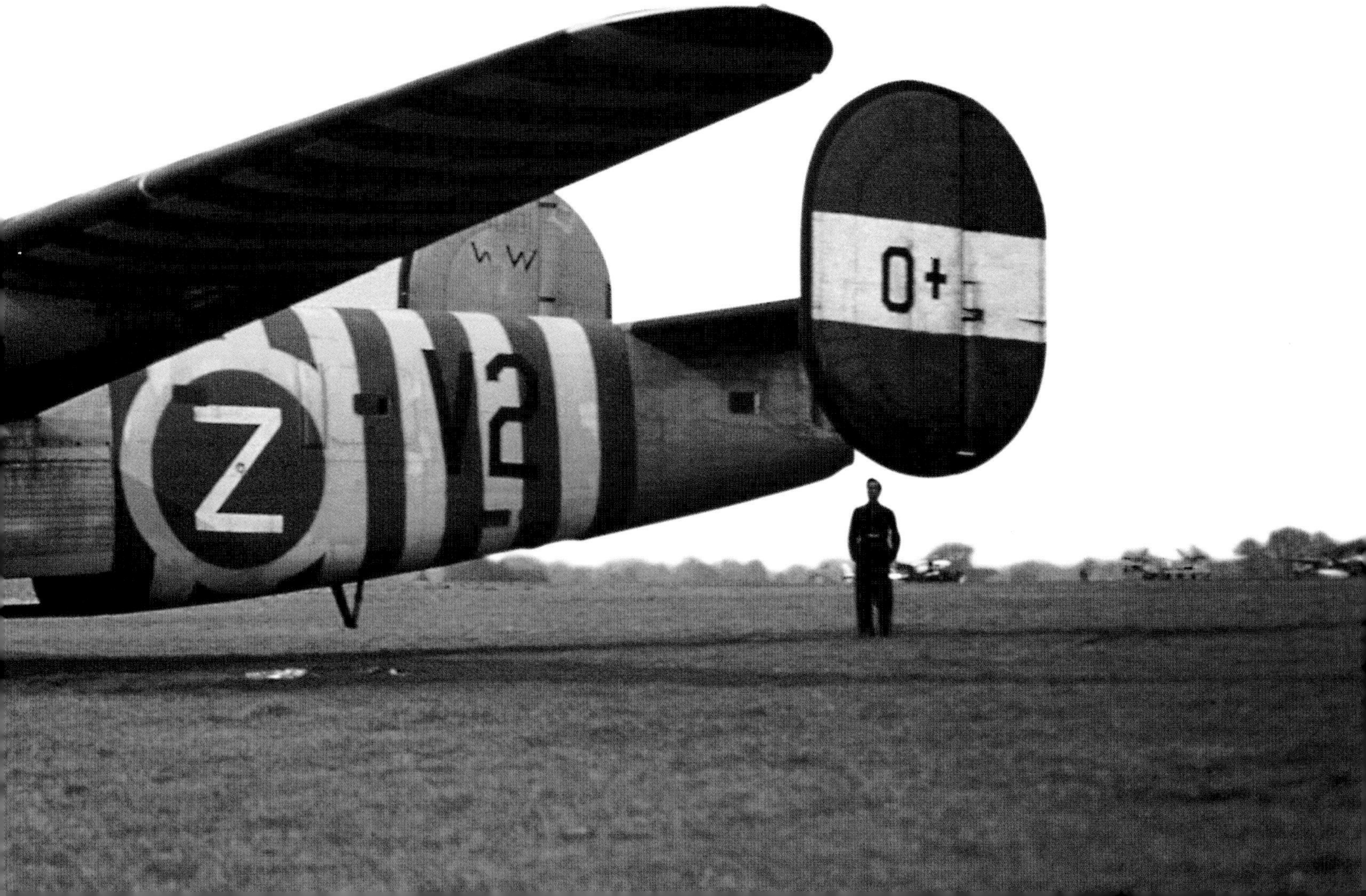

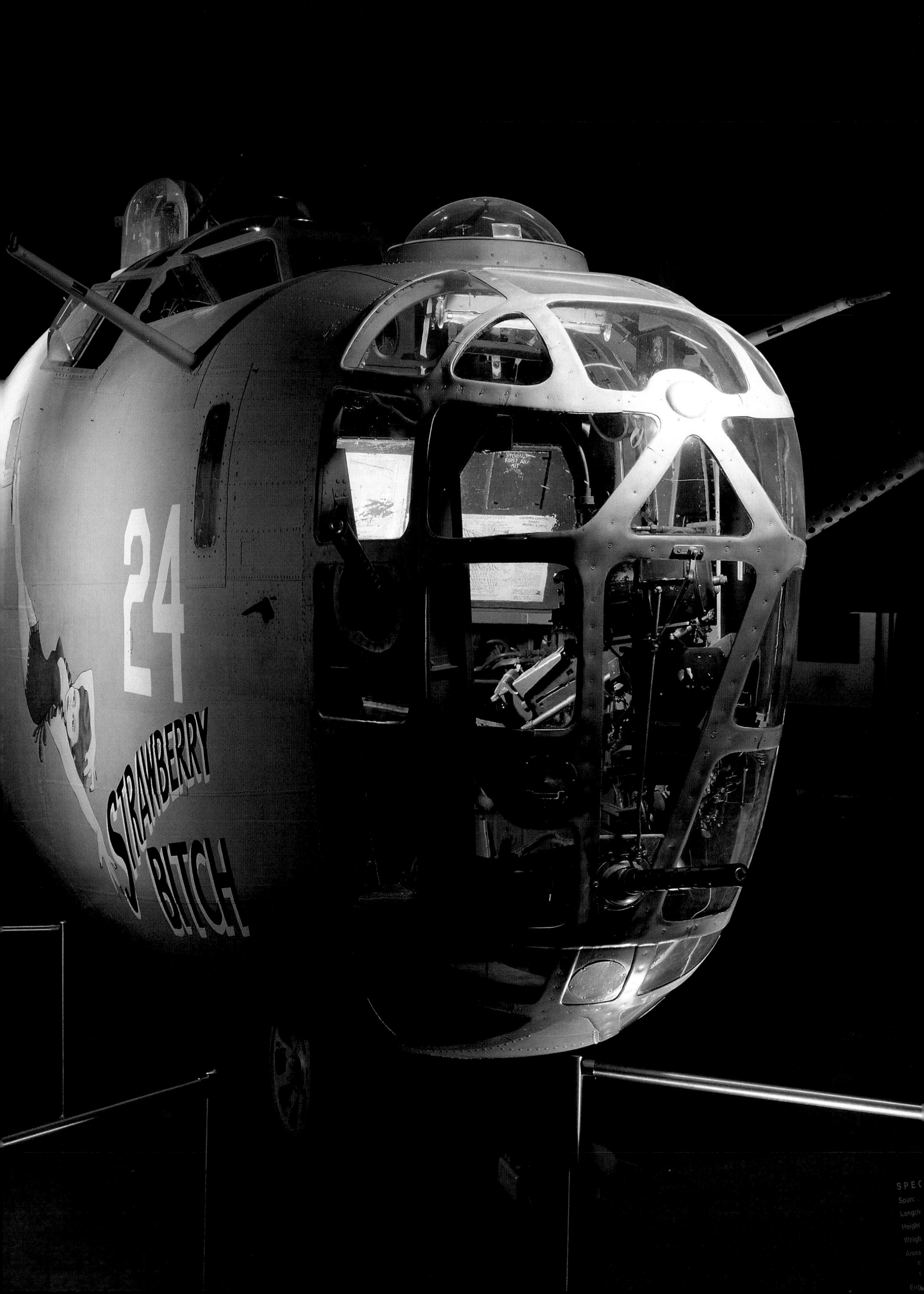
24
STRAWBERRY
BITCH

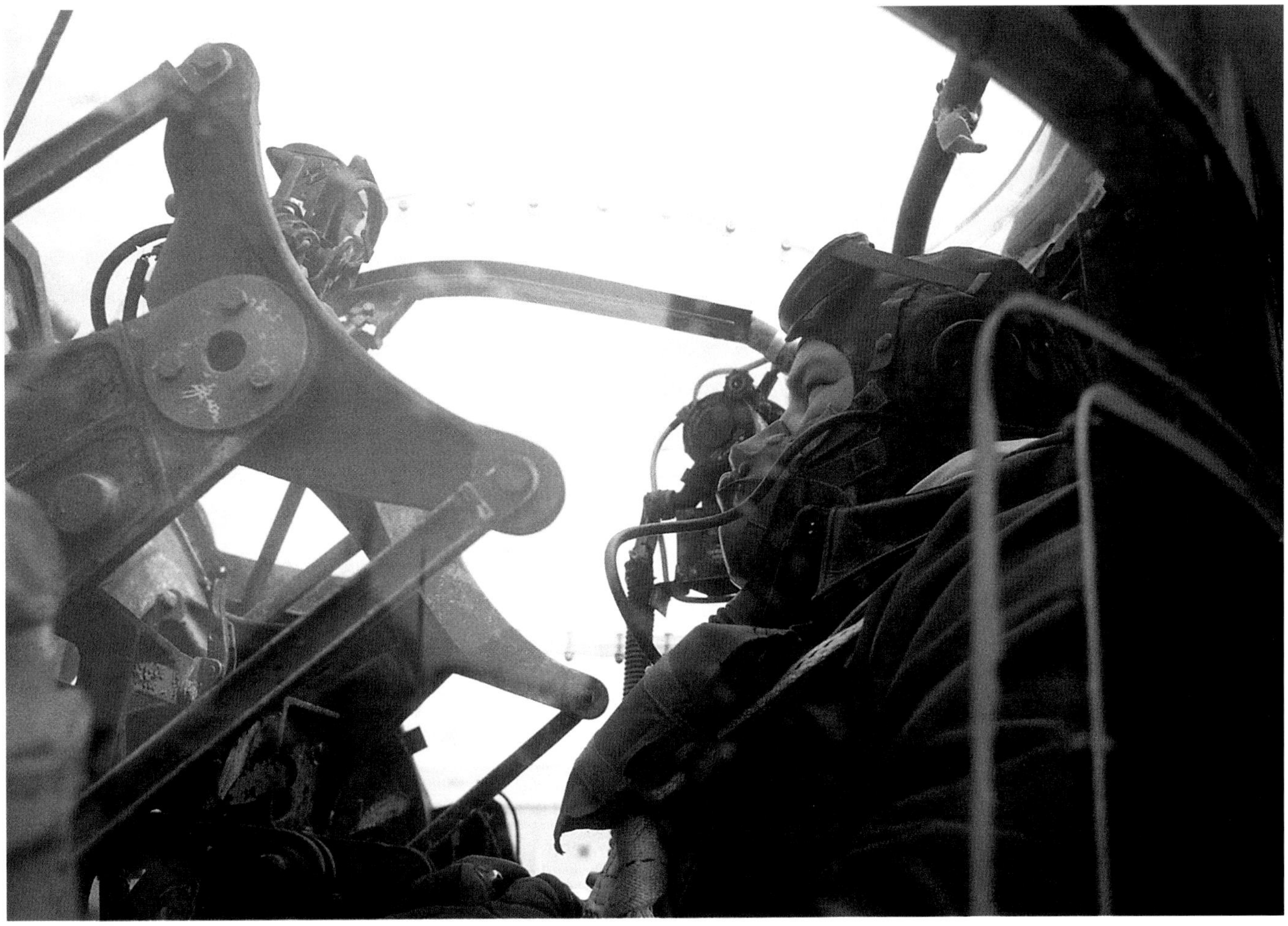

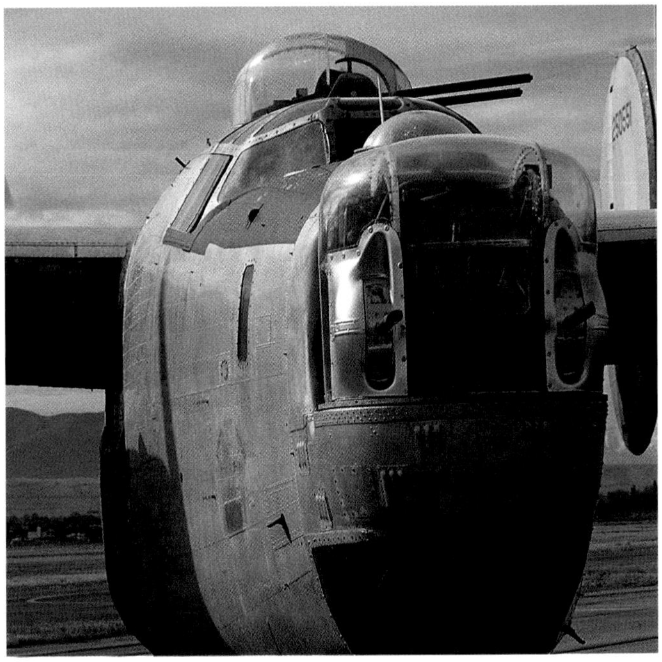

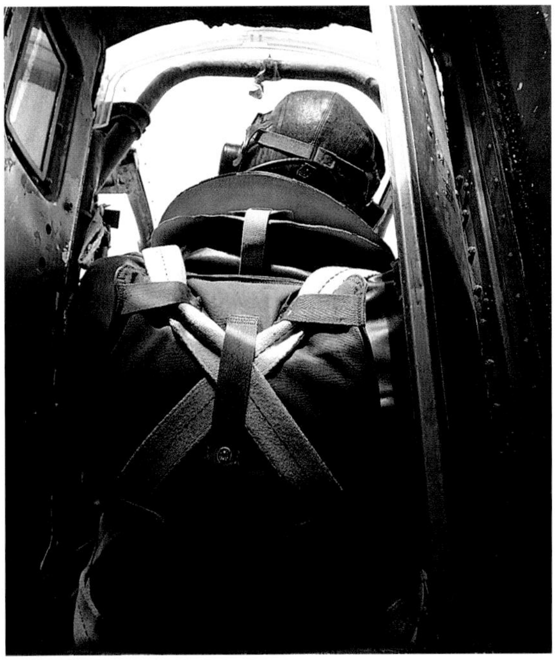

Left: The first significant version of the B-24 to see service with the USAAF was the D. The three hand-held .50s in the nose proved inadequate in the face of German and Japanese head-on fighter attacks.
USAF Museum, Dayton.

Top: The Emerson nose turret was lightly armored with bulletproof glass moving with the guns and 5/8 in and 3/8 in armor plate below.

Bottom left: The Emerson 127 was not a streamlined design and degraded the B-24's overall performance.

Bottom right: Space was at a premium in the Emerson nose turret. Whenever the gunner swiveled the turret, the icy slipstream would scream through the cracks, eliciting yells of protest from the rest of the crew.

"The Martin top turret was as good as or better than any of the others. We had an Emerson in the nose and a Consolidated turret in the tail. We didn't get the Sperry ball turret until we were ready to come home. I thought the Martin turret was the smoothest operating turret. There was not a lot of protection. Your head's sticking out there through the top, and the seat had no way of being armored. All the guts, the electrical stuff, was under the seat, so there was hardly any place to put anything. It was a tight fit with the ammo cans right in front of your knees, and right in front of that was a quarter inch of armor plate.

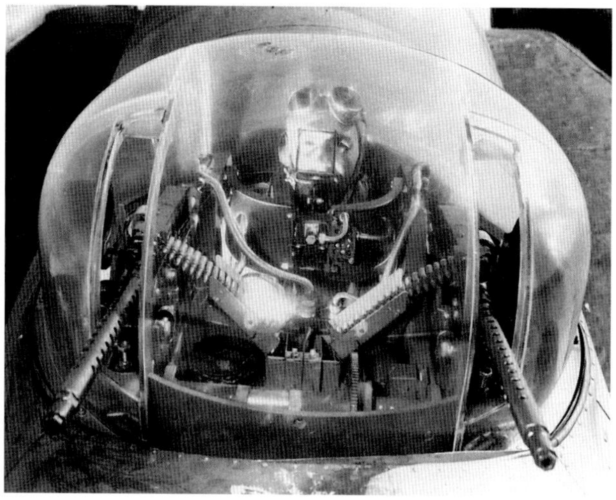

"Most of Japanese fighters used the same attack method as we did, the pursuit curve. They did scare the hell out of us a few times when they drove straight down through our formation. I don't really think they were actually kamikazes. These guys would just dive straight down through a formation with their guns wide open. It was extremely unnerving. But most them did the pursuit curve. My own feeling about it was, you just filled the sky full of lead and let them run into it."

JIM RODELLA, B-24 GUNNER, 64TH SQUADRON, 43RD BOMB GROUP (KENSMEN), 5TH AIR FORCE

ABOVE: THE MARTIN TOP TURRET WAS ONE OF THE MOST SUCCESSFUL AND EFFICIENT DESIGNS TO SEE SERVICE DURING THE SECOND WORLD WAR.
BELOW: A GUNNER LINES UP HIS TYPE N-6A GUNSIGHT IN A MARTIN TOP TURRET.

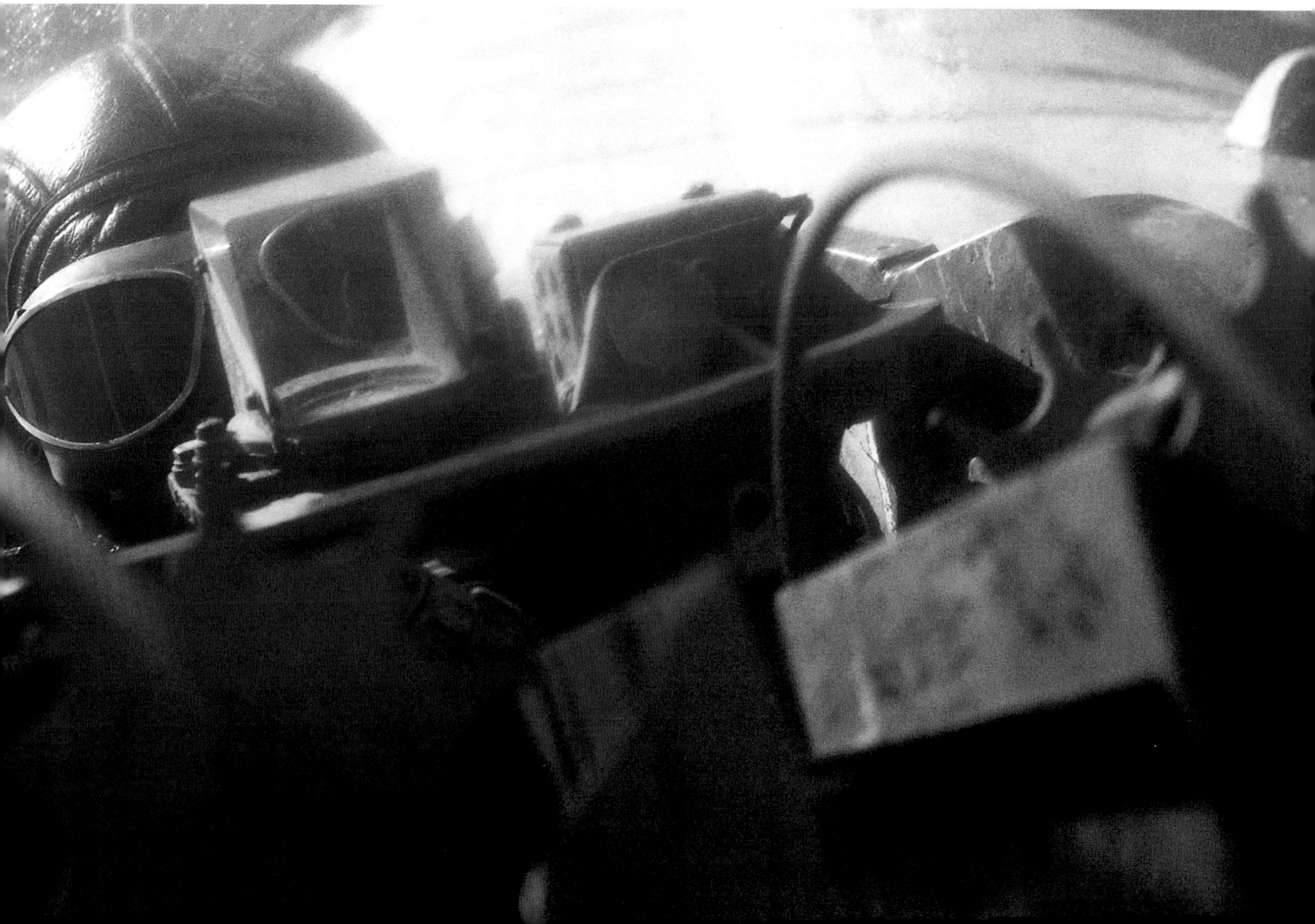

"To be honest with you, I thought the front turret was a bit of a bummer. You couldn't swing the turret because it would throw the airplane trim off. The pilot would say, 'Leave the goddamn turret alone.' The slipstream was also a problem. If you rotated the turret to either side, there were these rubber canvas jackets that were designed to keep the air out. But when you moved the turret, the slipstream would come howling through. The bombardier and navigator would scream like a son of a bitch because the wind coming through was as low as sixty degrees below zero.

"I preferred the waist gunner position, because the window was right there. If there was a problem, all you had to do was throw the gun aside and jump through the window. In the old D model, the escape hatch was between the waist and tail turret. Getting out of the tail wasn't easy. It was a more involved process, because your parachute was hanging up outside the turret. You couldn't wear your chute inside the tail turret because there was no room! We left the door open for that very reason. Getting out of the ball turret would have been a real problem.

I flew a couple of missions in the ball turret, and man, I'm telling you, that was hairy, really hairy.

"In the B-24, the ball turret was retractable. Before it was lowered, you had to roll it over with a hand crank so the guns were pointing straight down. After that, you stepped in, sat down and curled up. They would close the hatch and latch it. You have to remember that you're in a very small space. Your legs are up level with your head, your weight is on your back, and you're looking between your legs through the gunsight. Your neck is sort of bent and your back is rounded. There was a safety strap that you had to make damn sure was clipped on in case the door accidentally flew open. The trigger handles were located on the gunsight case. The left foot pedal controlled the intercom switch; the right foot pedal controlled the horizontal azimuth of the automatic gunsight. The one thing that was a real pain was when someone up front took a piss. It would splatter all over the turret and ice up the glass."

JOE TADDONIO, B-24 AIR GUNNER,
NOSE, WAIST, BALL AND TAIL TURRET

THE GUNNER'S VIEW FROM THE MARTIN TOP TURRET IN A B-24.

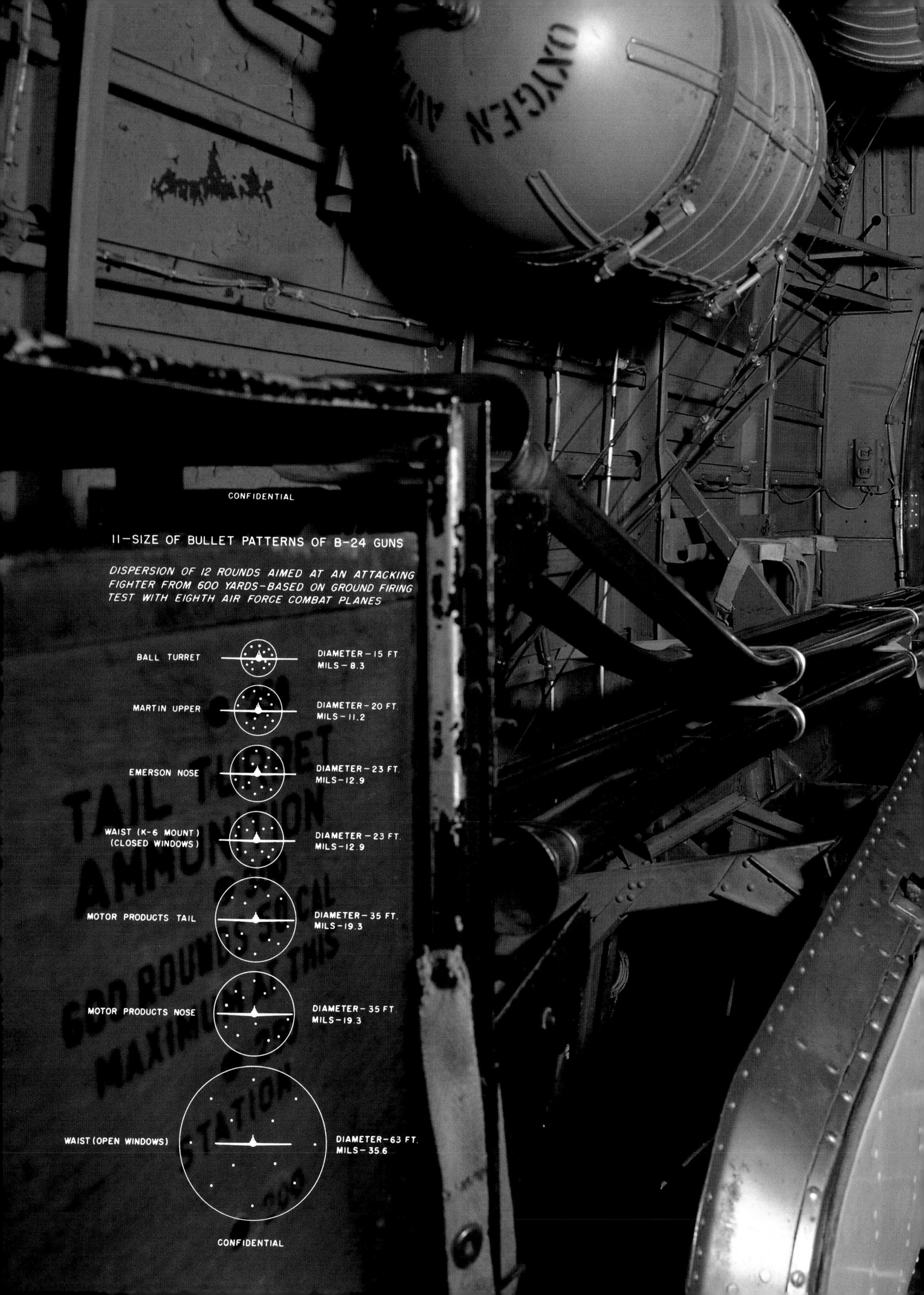
OXYGEN

CONFIDENTIAL

11—SIZE OF BULLET PATTERNS OF B-24 GUNS

DISPERSION OF 12 ROUNDS AIMED AT AN ATTACKING FIGHTER FROM 600 YARDS—BASED ON GROUND FIRING TEST WITH EIGHTH AIR FORCE COMBAT PLANES

BALL TURRET	DIAMETER—15 FT. MILS—8.3
MARTIN UPPER	DIAMETER—20 FT. MILS—11.2
EMERSON NOSE	DIAMETER—23 FT. MILS—12.9
WAIST (K-6 MOUNT) (CLOSED WINDOWS)	DIAMETER—23 FT. MILS—12.9
MOTOR PRODUCTS TAIL	DIAMETER—35 FT. MILS—19.3
MOTOR PRODUCTS NOSE	DIAMETER—35 FT. MILS—19.3
WAIST (OPEN WINDOWS)	DIAMETER—63 FT. MILS—35.6

CONFIDENTIAL

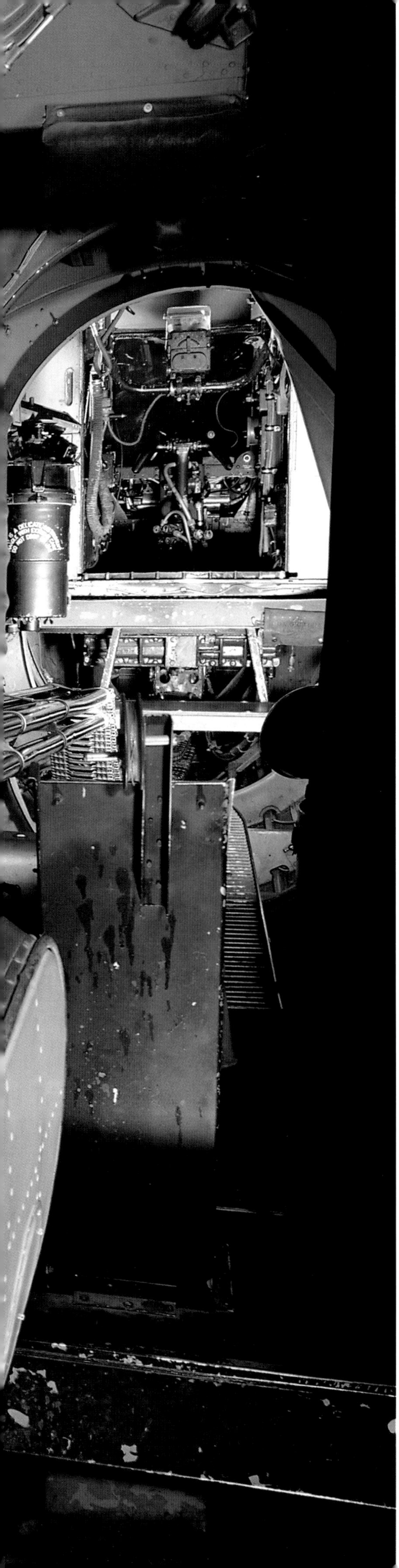

Section II

RESTRICTED
AN 11-45G-5

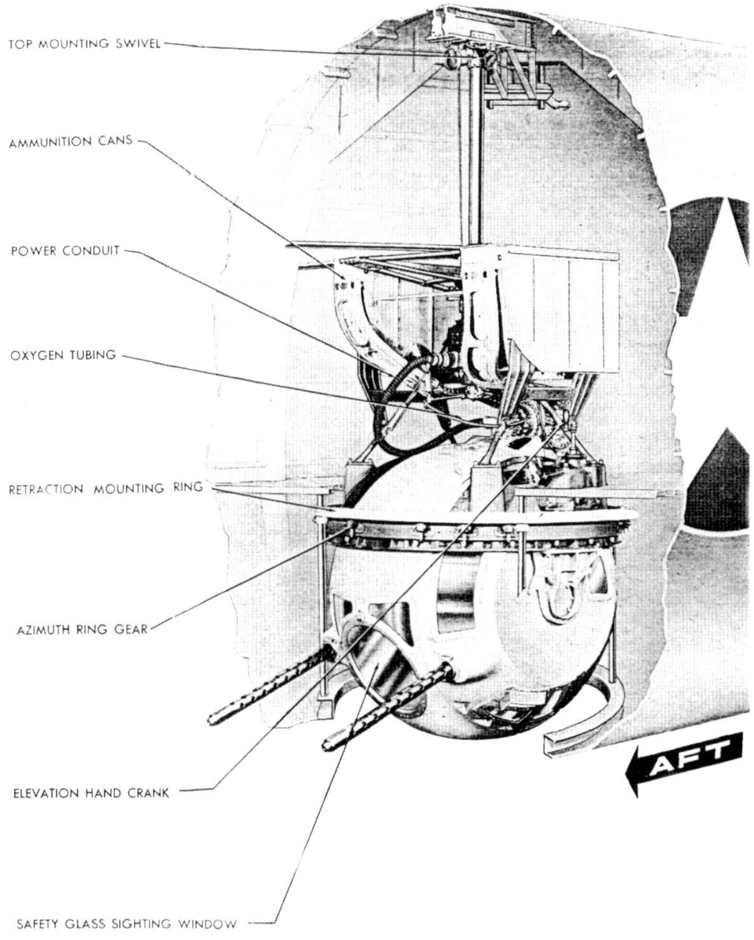
TOP MOUNTING SWIVEL
AMMUNITION CANS
POWER CONDUIT
OXYGEN TUBING
RETRACTION MOUNTING RING
AZIMUTH RING GEAR
ELEVATION HAND CRANK
SAFETY GLASS SIGHTING WINDOW
AFT

Figure 1—Type A-13A Turret (Retracted Position)

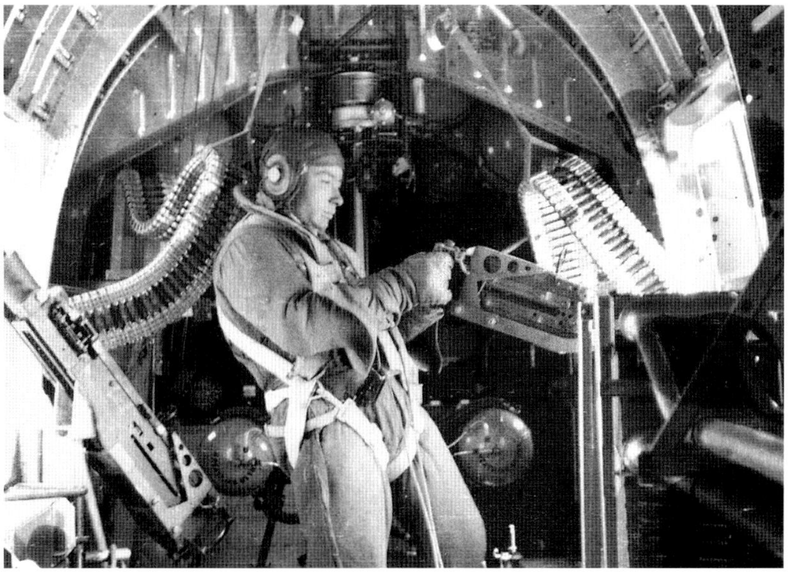

LEFT: FROM THE WAIST POSITION LOOKING BACK, ONE CAN CLEARLY SEE THE LARGE AMMUNITION BOXES (30 SECONDS WORTH OF FIRING) AND FEED CHUTES THAT SUPPLY THE CONSOLIDATED REAR TURRET. ENTRY IN AND OUT OF THE REAR TURRET WAS EXTREMELY DIFFICULT AND REQUIRED A GREAT AMOUNT OF AGILITY.
NATIONAL AVIATION MUSEUM, OTTAWA.
INSET: SIZE OF BULLET PATTERNS OF B-24 GUNS.

TOP: BECAUSE OF THE B-24'S LOW GROUND CLEARANCE, THE SPERRY BALL TURRET WAS FULLY RETRACTABLE.
BOTTOM: A SOLITARY B-24 WAIST GUNNER MANS HIS 65-POUND .50-CALIBER MACHINE GUN.

"The B-17 got all the glory, we did all the work!"

WALTER GENTRY,
FLIGHT ENGINEER / GUNNER,
460TH BOMB GROUP, 15TH AIR FORCE

"I found the Emerson front turret cramped, cold and noisy. I heard stories of people freezing their butts up there, but we had electrical suits that kept us warm. But I loved the front turret with those twin fifties. It was a real good piece of machinery, and I was pretty good at it too. I may have damaged an Fw 190 or 109 on occasion, but I didn't claim any kills.

"The German frontal attacks were such a fleeting thing. It was up and over before you knew it. With a closure rate of 600 miles per hour plus, they'd come in zippedy-doo-da! You were either hit or not. It was a real quick encounter."

2ND LIEUTENANT JOHN HUTTON,
B-24 NAVIGATOR / NOSE GUNNER,
763 SQUADRON, 460TH BOMB GROUP,
15TH AIR FORCE

A TAIL GUNNER TAKES AIM. THE TAIL GUNNER WAS PROTECTED BY A BULLETPROOF GLASS PANEL 2.2-IN (52-MM) THICK. DIRECTLY BELOW HIM WAS A 7/8-IN ARMOR PANEL, AND BESIDE EACH KNEE WERE TWO MORE SMALL PANELS EACH 3/8 IN THICK.
RIGHT: THE CONSOLIDATED TAIL TURRET WAS A VERY SUCCESSFUL DESIGN AND MANY WERE FITTED AS NOSE AND TAIL TURRETS ON THE B-24.
USAF MUSEUM, DAYTON.

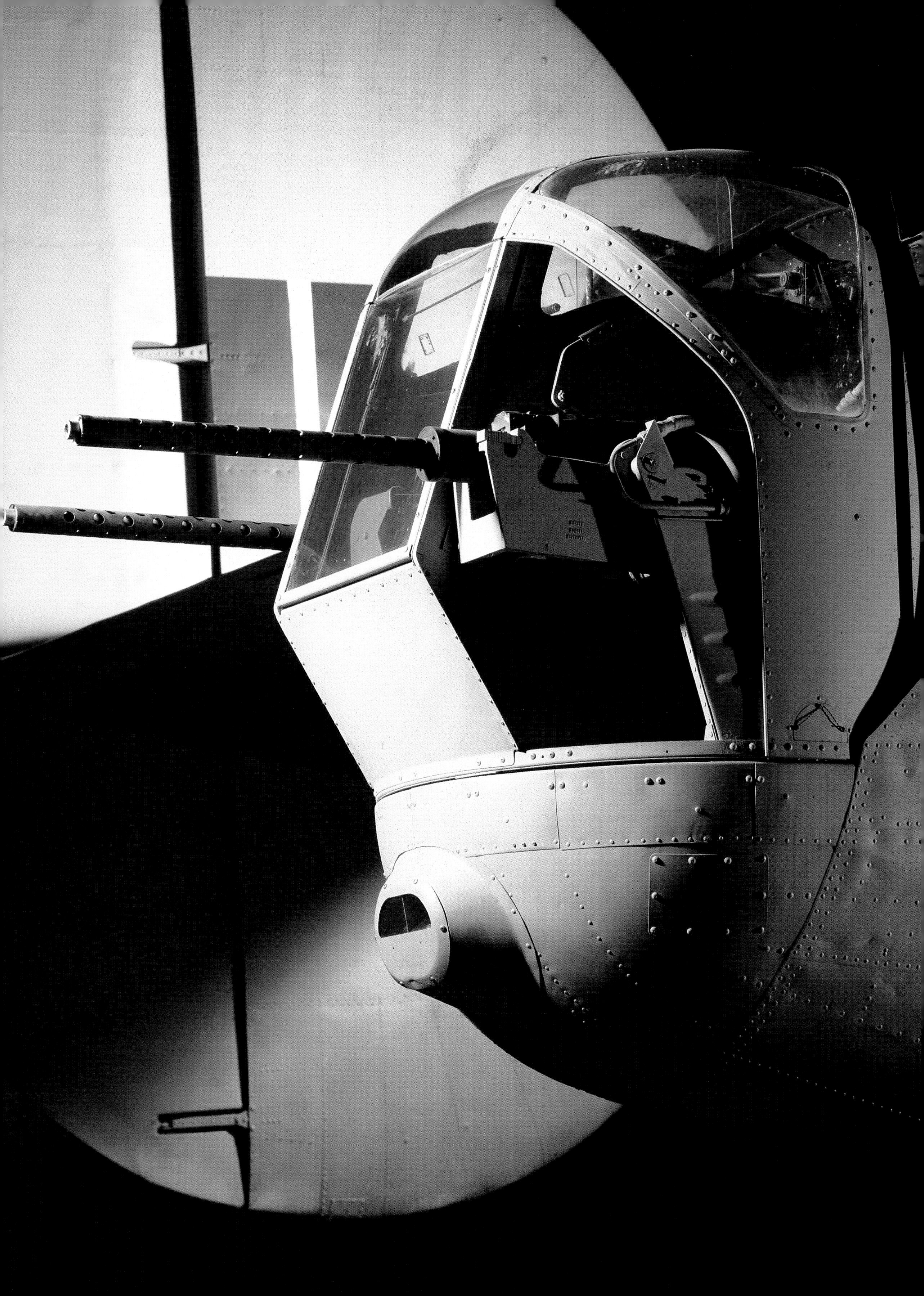

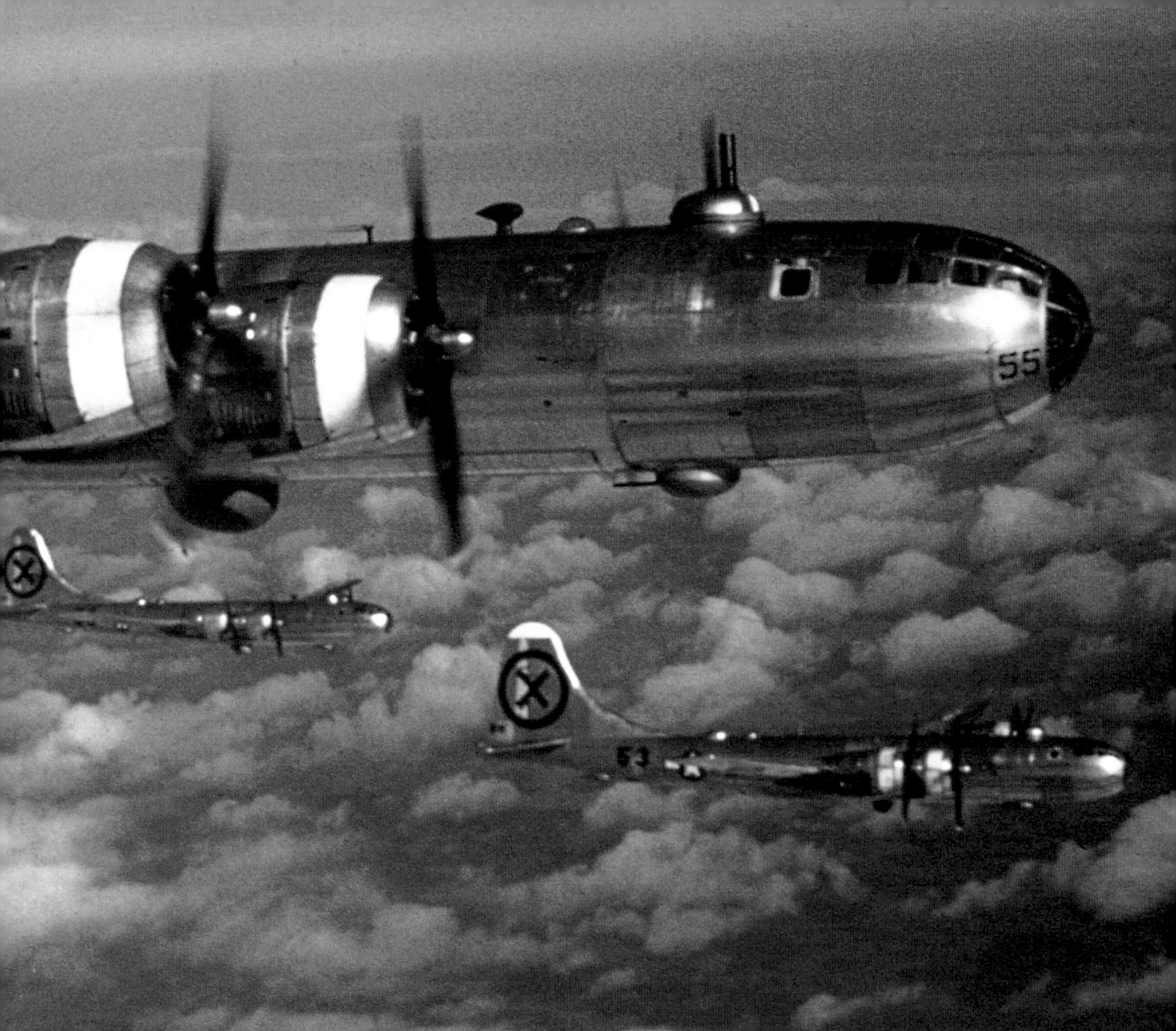
55
53
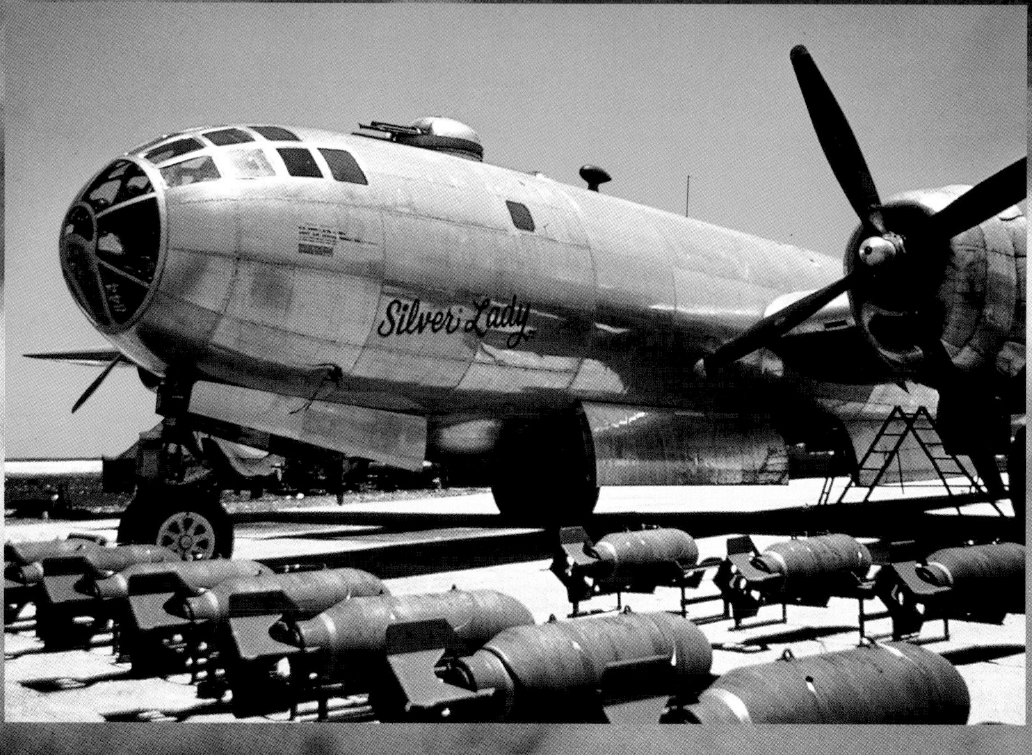
Silver Lady

Boeing B-29
Superfortress

Even with the world's most advanced computer-aided fire-control system, the turret-armed B-29 was, like all bombers, vulnerable to determined attacks by heavily armed fighters.

When the B-29 first entered service in 1943, it was the most advanced bomber in the world. It was a quantum leap over the B-17 and B-24. Fully pressurized, it could fly over 3,000 miles (5,230 km) with a bomb load of up to 20,000 lb (9,072 kg). But one of its most unique features was the General Electric fully remote-controlled turret (RCT) system. No longer did gunners have to wrap themselves in electric suits and push themselves into cold, cramped turrets. The B-29 was equipped with four remote-controlled turrets and a gunner in the tail position. The other gunners operated their turrets from a pressurized position just aft of the wing. There, the upper and two side blister gunners controlled the rear upper, rear lower and front lower turret. The navigator / nose gunner controlled the front top and front lower turret. The side blister gunners could also operate the tail turret. The tail turret gunner could operate only his own guns.

The B-29 was armed with twelve .50-caliber Colt Browning machine guns. The upper forward turret carried four guns, while the other turrets contained two. At the heart of the General Electric–designed system were five central fire-control computers housed in armored compartments beneath the cabin floors. As the gunner aligned his sights, the computer would analyze the information taken from the gunsight (parallax, ballistics, deflection) and send electrical impulses to the engaged turret. These signals were then amplified and fed directly into an Amplidyne generator; there they energized the turret drive-motors to the required aiming point, directed by the gunner. The navigator operated a handset by which he keyed into the computer other information such as outside air temperature, barometric altitude and indicated air speed. Provided the gunner tracked the target smoothly with his sight and adjusted the range, vital allowances for aiming normally associated with manned turrets were taken care of by the computers.

LEFT: B-29S FROM 9TH BOMB GROUP HEAD FOR THEIR TARGET.
THE UPPER FORWARD TURRET IS MADE READY FOR VERTICAL DIVING ATTACKS BY JAPANESE FIGHTERS.

INSET: THE UPPER FORWARD TURRET ON THE B-29 CARRIED FOUR .50-CALIBER MACHINE GUNS.
BASED IN THE MARIANAS, SILVER LADY STANDS READY TO RECEIVE HER BOMB LOAD.

ABOVE: THE TOP GUNNER / CENTRAL FIRE CONTROLLER WAS RESPONSIBLE FOR
DIRECTING AND COORDINATING THE FIRING OF THE OTHER TURRETS.

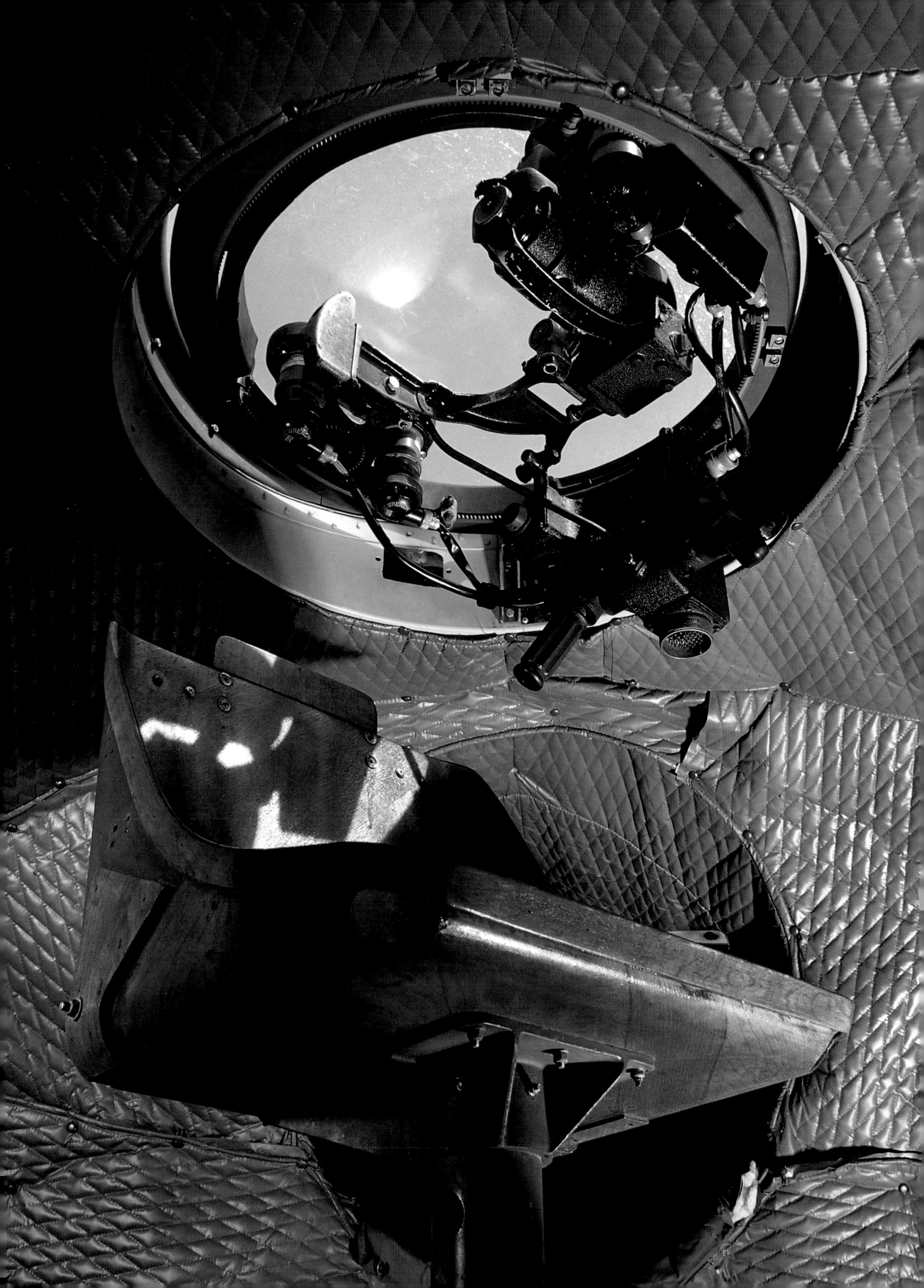

One of the B-29's most impressive features was that it allowed all the gunners to control two turrets from a single gunsight — all it took was a flick of a switch. Gunners could then double their firepower when needed. The top gunner — the Central Fire Controller — was given control of the system. It was his job was to co-ordinate the firing and assign targets to the other gunners.

As effective as this system was, the fact remained that many B-29s were shot down by Japanese fighters. Throughout the Second World War, the fighter always had the advantage over the bomber, no matter how sophisticated the defensive armament was. With the introduction of jet fighters at the conclusion of the war, the days of the multi-turret armed bomber came to an end.

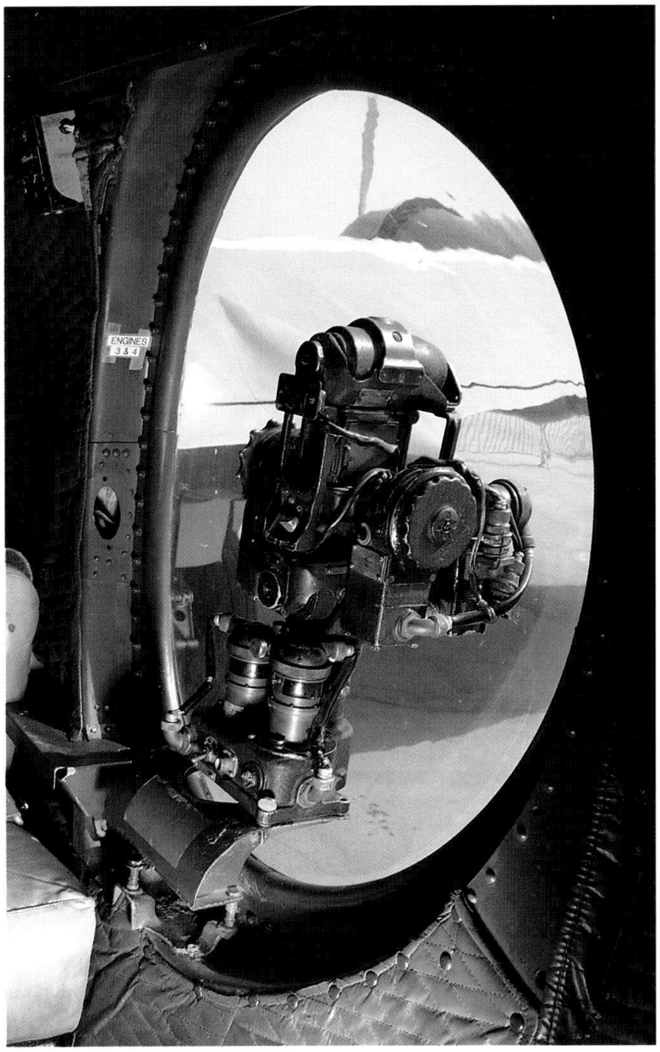
ENGINES
3 & 4

"I was assigned to the tail turret in the B-29. Compared to the ball turret position I had trained in, the tail position was spacious. It was two foot square and about six feet high. You could stand up in it. I remember crawling back, opening the door, getting in, standing up and looking around. The view in three directions was great and the escape hatch was close at hand. The tail position also had a folding seat that slid up behind you. When you pulled it down and opened it up, the life raft was attached to form the cushion.

"The second time I encountered a fighter was the time we were jumped coming out of Osaka. It was a twin-engine fighter, and I know darn well he was trying to get us by surprise. To this day, I shudder to think what might have happened had he been able to catch us napping. We were unwinding, because once you got a hundred miles off the coast you began begin to relax. There was a cloud base down below us, and for some reason I was still watching when I saw this guy pop up and drop back down. The first thing I know, he overhauled us down through the clouds and then popped up behind us. I alerted everyone. I was able to open fire when he swung across the back of me. He was firing, I was firing, and I can still see my bullets smashing into his engine and wing, with parts of the engine and wing flying off. I could see that my bullets were hitting his right engine, but I couldn't force myself to swing the guns over to the pilot's compartment. We were told to trust the sight and not the tracers. So I did, of course. If I had shot at the pilot's compartment, I might have killed some poor guy and I'm glad now I didn't."

Sergeant Andy Doty, B-29 Tail Gunner,
19th Bomb Group, Guam

"When you fired those two big .50-calibers, it's really something. They rattle you all over the place with spent shell casings flying back, pinging off the floor. It's pretty exciting.

"The tail gunner's position was very isolated. You're cut off from the rest of the crew, except for the intercom, of course. It was a long way to the next gunner's position."

Bob Blumberg, B-29 Tail Gunner,
73rd Bomb Group

Left: The B-29 Central Fire Controller's seat.
Directly above him is the upper ring sighting station.
Confederate Air Force.

Above: Close-up of the gunsight in the B-29's side blister.
Confederate Air Force.

"April 7th was the wildest fighter attack. I got credit for two kills on that trip. The first one I saw coming. As he came in, I figured out that I could open up on him quite a lot further out than my gunsight indicated. Somebody up front was screaming, 'Get that fighter in front.' I opened up on him at about 1,200 yards. The crew said he started losing pieces as soon as I opened up on him. By the time he got to us, he bailed out. The other Japanese fighter was a little high and slightly behind us. That was more a reflex action. I had seen several fighters come in and appear to stop momentarily. This fighter looked like he was trying to make a hard turn and come around to make an attack or follow us. Our central fire-control system required smooth tracking and framing of the target to be accurate. By anticipating where he would stop, my guns were with my sight. I had my guns set and waited, because I knew he was going to stop momentarily. When he did, I let him have it. At the time, I thought somebody else was shooting at him, but my two turrets were converging on him."

BILL AGEE, B-29 AIR GUNNER,
73RD BOMB GROUP

ABOVE: THE *ENOLA GAY* TAIL TURRET. THE TAIL TURRET WAS THE ONLY GUNNER'S POSITION WHERE HE WAS NOT REMOTE FROM HIS GUNS. BOTH SIDE BLISTER GUNNERS COULD TAKE CONTROL OF THE TAIL TURRET IN CASE OF EMERGENCY.
NATIONAL AIR AND SPACE MUSEUM.

RIGHT: CLOSE-UP OF THE *ENOLA GAY*'S TAIL GUN FIRE CONTROLLER INCORPORATING A REFLECTOR-TYPE SIGHT. AS THE GUNNER MANIPULATED THE TWO KNOBS AND ALIGNED THE SIGHT ONTO HIS TARGET, A CENTRAL FIRE CONTROL COMPUTER ANALYZED THE SIGNALS AND SENT ELECTRICAL IMPULSES TO THE MOVE THE GUNS.
NATIONAL AIR AND SPACE MUSEUM.

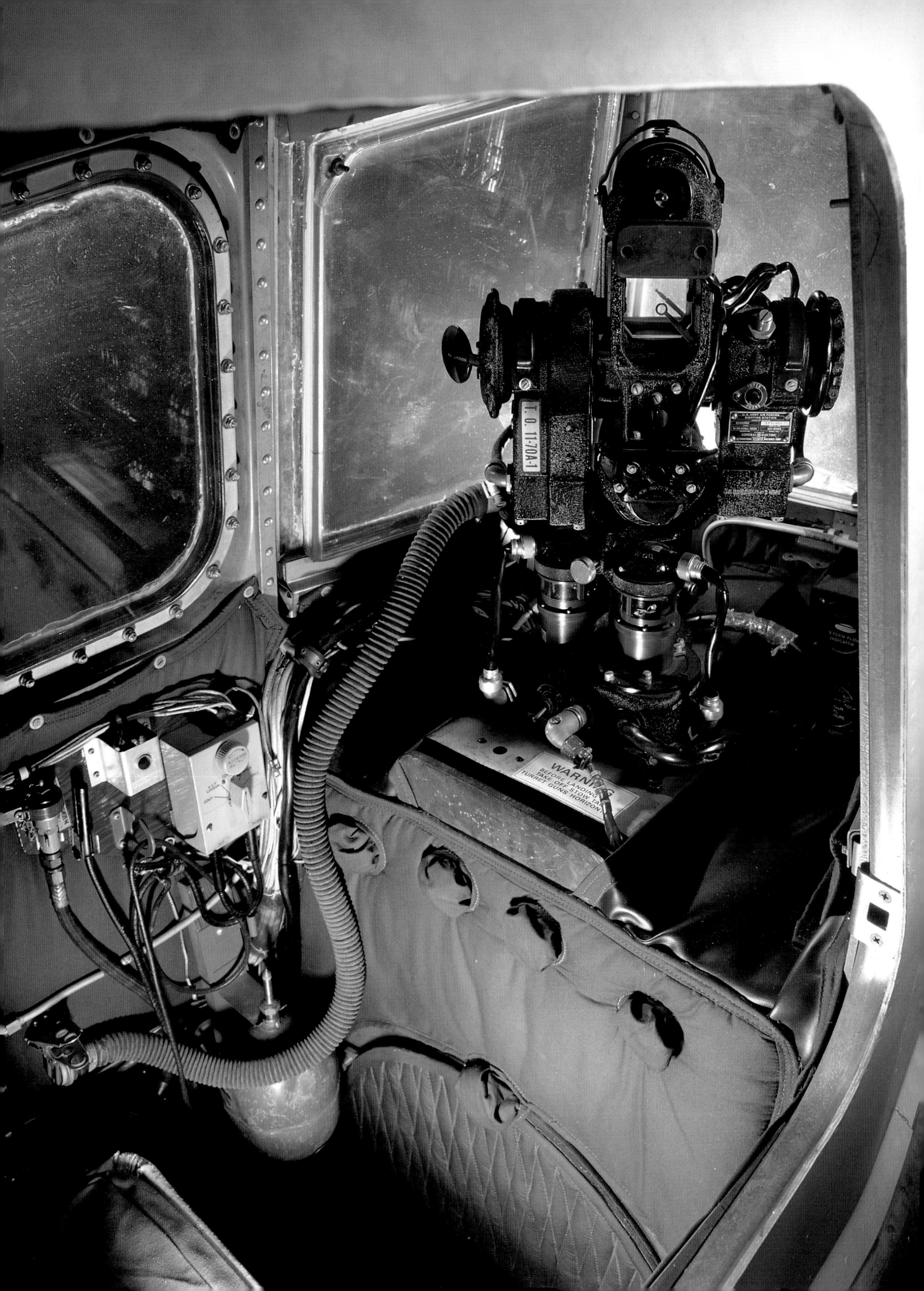
T. O. 11-70A-1

AVRO
LANCASTER

UNDER THE COVER OF DARKNESS, LANCASTER GUNNERS FOUGHT A VERY DIFFERENT WAR. IT WAS A WAR WHERE THE ENEMY HID IN THE SHADOWS AND CONSTANT VIGILANCE OFTEN MEANT THE DIFFERENCE BETWEEN LIFE AND DEATH.

Bomber Command's premiere bomber, the Avro Lancaster, was a true heavyweight. Capable of lifting a single 22,000 Grand Slam bomb, the Lancaster was, when compared to its American counterparts, one of the more lightly armed four-engine bombers of the war.

The Lancaster crew consisted of a pilot, flight engineer, navigator, wireless operator, bomb aimer, mid-upper gunner and rear gunner. The bomb aimer also manned the front gun turret. The Lancaster's original armament consisted of ten .303-caliber Browning machine guns mounted in four powered turrets. The Lancaster was fully equipped with Frazer-Nash (FN) turrets with two . 303-caliber machine guns mounted in the FN 5 nose turret, two in the FN 50 mid-upper, two in the FN 64 mid-under, and four in the FN 20 tail turret. The FN 64 mid-under did not see widespread use, and only four squadrons were equipped with this turret. A periscopic sight was used for sighting but proved ineffective in acquiring a target. This led to the decision to cancel its installation on production aircraft. Canadian-built Lancasters were equipped with a Martin mid-upper turret armed with two .50-caliber machine guns, and in June 1944 the Rose Rice rear turret was used operationally for the first time on British-built Lancasters. The Rose turret was armed with two .50-caliber machine guns and had a clear cutaway panel that offered an exceptional view. The cutaway panel also provided the gunner with a quick means of escape. All the gunner had to do was push himself through the opening and he was free of the aircraft. The new turrets proved very popular and were a welcome, if belated, addition to the Lancaster's rear defense. Toward the end of the war, some Lancasters were equipped with AGLT (Automatic Gun-Laying Turret radar) and two .50-caliber machine guns. The AGLT was not perfect, since it depended on a perfectly discriminating

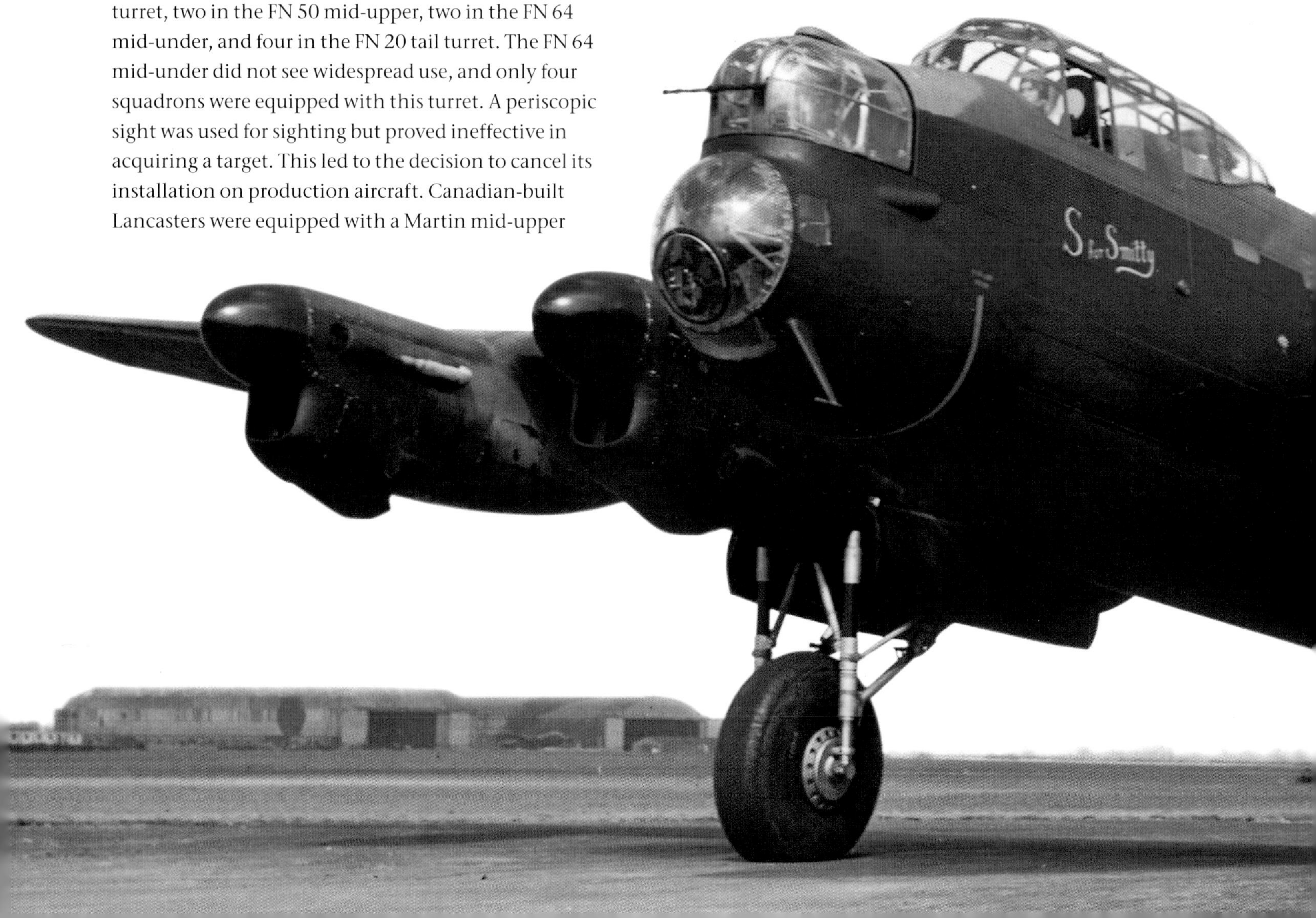
S for Smitty

IFF (Identification Friend or Foe). Even with 99 percent IFF accuracy, it was estimated that gunners would shoot down four times as many bombers as night fighters.

"I fired my guns four times at what I thought were German fighters. I put bullets right through them, but nothing happened. That's the way it appeared to me. My bullets were ineffective because of the range and because the German night fighters were armed with cannons. They could out-gun us every time. It became a game of who saw who first, and it was certainly demoralizing to have four guns, each shooting 1,100 rounds a minute, and still not be effective.

"Going into the target, you would become very apprehensive and nervous. The searchlights, flak, the fighters, the danger of collision, the danger of getting hit with your own bombs. If you had time to stop and think about being frightened, you weren't doing your job. Especially when you're flying at night for hours and hours in total darkness and then when you got over the target it was lighter than day. All hell was breaking loose. The Germans knew that we dropped TIs (target indicators) and we all had to fly over those indicators at a designated height at a designated time. The Germans knew this, so they just fired their flak through the bloody TI. The only thing they didn't know was our bombing height."

WARRANT OFFICER SECOND CLASS FRED VINCENT,
AIR GUNNER, LANCASTER TAIL TURRET,
189 SQUADRON, 5 GROUP, RAF

"I found the rear turret in the Lancaster, as compared to the Halifax, to be a much cleaner design. There were large panes of perspex in the Frazer Nash, whereas in the Boulton Paul turret, you had all those individual panes. "The only thing that really bugged me about being in the rear turret was being confined in one position for so long. What I used to do was open the doors and lie back on the slide and stretch my back. Of course, I told the mid-upper before I did my stretch.

"On this one trip, I kept seeing what looked like shooting stars, and at the time, the Germans were always shooting up 'scarecrows' and stuff to scare the hell out of you. You never knew what the heck it was, and I saw these blue streaks going off and mentioned them to the mid-upper gunner. He said, 'Yeah, I've been watching them.' We found out a couple of days later that they were Messerschmitt Me 163 rocket fighters."

FLYING OFFICER BILL COLE, AIR GUNNER,
HALIFAX AND LANCASTER TAIL TURRET,
428 SQUADRON, 6 GROUP, RCAF

AN RCAF LANCASTER WARMS UP BEFORE TAKE-OFF.

"The Frazer Nash top turret was a very good turret as far as all-round view was concerned. You had a very good picture of the whole sky all around you. The one thing I didn't like about the turret was getting in and out. With a heated flying suit, Mae West and parachute harness all strapped on, it was a little tough to get in and out of. The Lancaster was a difficult aircraft when it came time to bail out. When you compared the three — the Lanc, the Stirling and the Halifax — the Lanc was the toughest of them all.

"My first operation was to the Ruhr Valley, and I was scared stiff. I remember it like it was yesterday. We were sitting at 19,500 feet when I happened to look and saw another Lancaster coming over us with the bomb doors open. I was awestruck. I just watched as he drifted over us, hoping we wouldn't get hit with a 500-pound bomb.

"Our skipper didn't want us to fire on German night fighters. He wasn't one of the ones that used to fly straight and level. We used to do a little weaving and banking. This would allow us to look underneath the aircraft. If you've read any of Martin Middlebrook's books, you'll know there's a lot of them that didn't do any weaving, and they wound up getting shot down. Our skipper gave us a little bank every so often so we could look underneath the aircraft. He said, 'Keep your nose clean unless necessary and we'll stay out of harm's way,' which is exactly what happened. I must admit that we had a Cracker Jack crew. We were always on time. We were always where we were supposed to be, flight and height and everything else."

AL PARISANI, AIR GUNNER,
LANCASTER MID-UPPER, 106 SQUADRON, 5 GROUP, RAF

ABOVE: THE FRAZER NASH FN 50 TOP TURRET WAS DESCRIBED AS ROOMY AND COMFORTABLE BUT DIFFICULT TO EXIT IN CASE OF AN EMERGENCY. IT HAD AN EXCELLENT FIELD OF VIEW, AND ARMOR PROTECTION CONSISTED OF A 7-MM APRON AROUND THE FRONT OF THE TURRET.
BATTLE OF BRITAIN MEMORIAL FLIGHT.
RIGHT: THIS RAF AIR DIAGRAM IS A CUTAWAY DRAWING OF THE FN 50 TOP TURRET.

RESTRICTED
(For official use only)

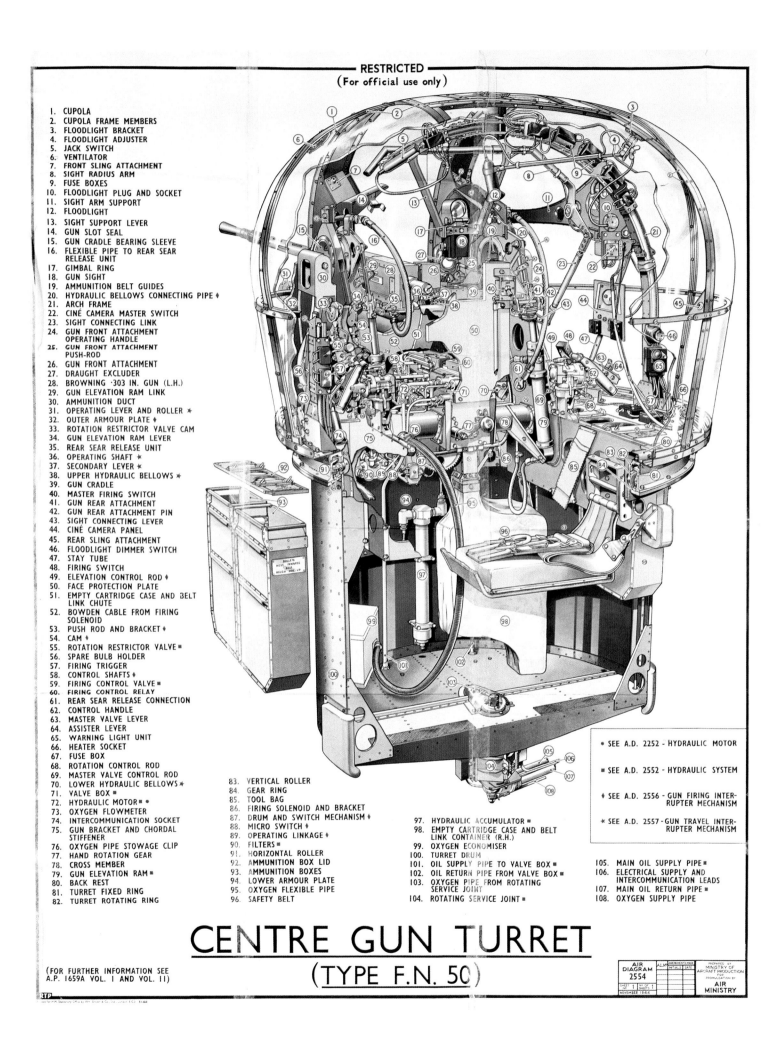

1. CUPOLA
2. CUPOLA FRAME MEMBERS
3. FLOODLIGHT BRACKET
4. FLOODLIGHT ADJUSTER
5. JACK SWITCH
6. VENTILATOR
7. FRONT SLING ATTACHMENT
8. SIGHT RADIUS ARM
9. FUSE BOXES
10. FLOODLIGHT PLUG AND SOCKET
11. SIGHT ARM SUPPORT
12. FLOODLIGHT
13. SIGHT SUPPORT LEVER
14. GUN SLOT SEAL
15. GUN CRADLE BEARING SLEEVE
16. FLEXIBLE PIPE TO REAR SEAR RELEASE UNIT
17. GIMBAL RING
18. GUN SIGHT
19. AMMUNITION BELT GUIDES
20. HYDRAULIC BELLOWS CONNECTING PIPE ♦
21. ARCH FRAME
22. CINÉ CAMERA MASTER SWITCH
23. SIGHT CONNECTING LINK
24. GUN FRONT ATTACHMENT OPERATING HANDLE
25. GUN FRONT ATTACHMENT PUSH-ROD
26. GUN FRONT ATTACHMENT
27. DRAUGHT EXCLUDER
28. BROWNING ·303 IN. GUN (L.H.)
29. GUN ELEVATION RAM LINK
30. AMMUNITION DUCT
31. OPERATING LEVER AND ROLLER ✱
32. OUTER ARMOUR PLATE ♦
33. ROTATION RESTRICTOR VALVE CAM
34. GUN ELEVATION RAM LEVER
35. REAR SEAR RELEASE UNIT
36. OPERATING SHAFT ✱
37. SECONDARY LEVER ✱
38. UPPER HYDRAULIC BELLOWS ✱
39. GUN CRADLE
40. MASTER FIRING SWITCH
41. GUN REAR ATTACHMENT
42. GUN REAR ATTACHMENT PIN
43. SIGHT CONNECTING LEVER
44. CINÉ CAMERA PANEL
45. REAR SLING ATTACHMENT
46. FLOODLIGHT DIMMER SWITCH
47. STAY TUBE
48. FIRING SWITCH
49. ELEVATION CONTROL ROD ♦
50. FACE PROTECTION PLATE
51. EMPTY CARTRIDGE CASE AND BELT LINK CHUTE
52. BOWDEN CABLE FROM FIRING SOLENOID
53. PUSH ROD AND BRACKET ♦
54. CAM ♦
55. ROTATION RESTRICTOR VALVE ■
56. SPARE BULB HOLDER
57. FIRING TRIGGER
58. CONTROL SHAFTS ♦
59. FIRING CONTROL VALVE ■
60. FIRING CONTROL RELAY
61. REAR SEAR RELEASE CONNECTION
62. CONTROL HANDLE
63. MASTER VALVE LEVER
64. ASSISTER LEVER
65. WARNING LIGHT UNIT
66. HEATER SOCKET
67. FUSE BOX
68. ROTATION CONTROL ROD
69. MASTER VALVE CONTROL ROD
70. LOWER HYDRAULIC BELLOWS ✱
71. VALVE BOX ■
72. HYDRAULIC MOTOR ■ •
73. OXYGEN FLOWMETER
74. INTERCOMMUNICATION SOCKET
75. GUN BRACKET AND CHORDAL STIFFENER
76. OXYGEN PIPE STOWAGE CLIP
77. HAND ROTATION GEAR
78. CROSS MEMBER
79. GUN ELEVATION RAM ■
80. BACK REST
81. TURRET FIXED RING
82. TURRET ROTATING RING

83. VERTICAL ROLLER
84. GEAR RING
85. TOOL BAG
86. FIRING SOLENOID AND BRACKET
87. DRUM AND SWITCH MECHANISM ♦
88. MICRO SWITCH ♦
89. OPERATING LINKAGE ♦
90. FILTERS ■
91. HORIZONTAL ROLLER
92. AMMUNITION BOX LID
93. AMMUNITION BOXES
94. LOWER ARMOUR PLATE
95. OXYGEN FLEXIBLE PIPE
96. SAFETY BELT
97. HYDRAULIC ACCUMULATOR ■
98. EMPTY CARTRIDGE CASE AND BELT LINK CONTAINER (R.H.)
99. OXYGEN ECONOMISER
100. TURRET DRUM
101. OIL SUPPLY PIPE TO VALVE BOX ■
102. OIL RETURN PIPE FROM VALVE BOX ■
103. OXYGEN PIPE FROM ROTATING SERVICE JOINT
104. ROTATING SERVICE JOINT ■
105. MAIN OIL SUPPLY PIPE ■
106. ELECTRICAL SUPPLY AND INTERCOMMUNICATION LEADS
107. MAIN OIL RETURN PIPE ■
108. OXYGEN SUPPLY PIPE

• SEE A.D. 2252 - HYDRAULIC MOTOR

■ SEE A.D. 2552 - HYDRAULIC SYSTEM

♦ SEE A.D. 2556 - GUN FIRING INTERRUPTER MECHANISM

✱ SEE A.D. 2557 - GUN TRAVEL INTERRUPTER MECHANISM

CENTRE GUN TURRET
(TYPE F.N. 50)

(FOR FURTHER INFORMATION SEE A.P. 1659A VOL. I AND VOL. II)

AIR DIAGRAM 2554 — SHEET No 1 — No OF SHEETS 1 — NOVEMBER 1944 — PREPARED BY MINISTRY OF AIRCRAFT PRODUCTION FOR PROMULGATION BY AIR MINISTRY

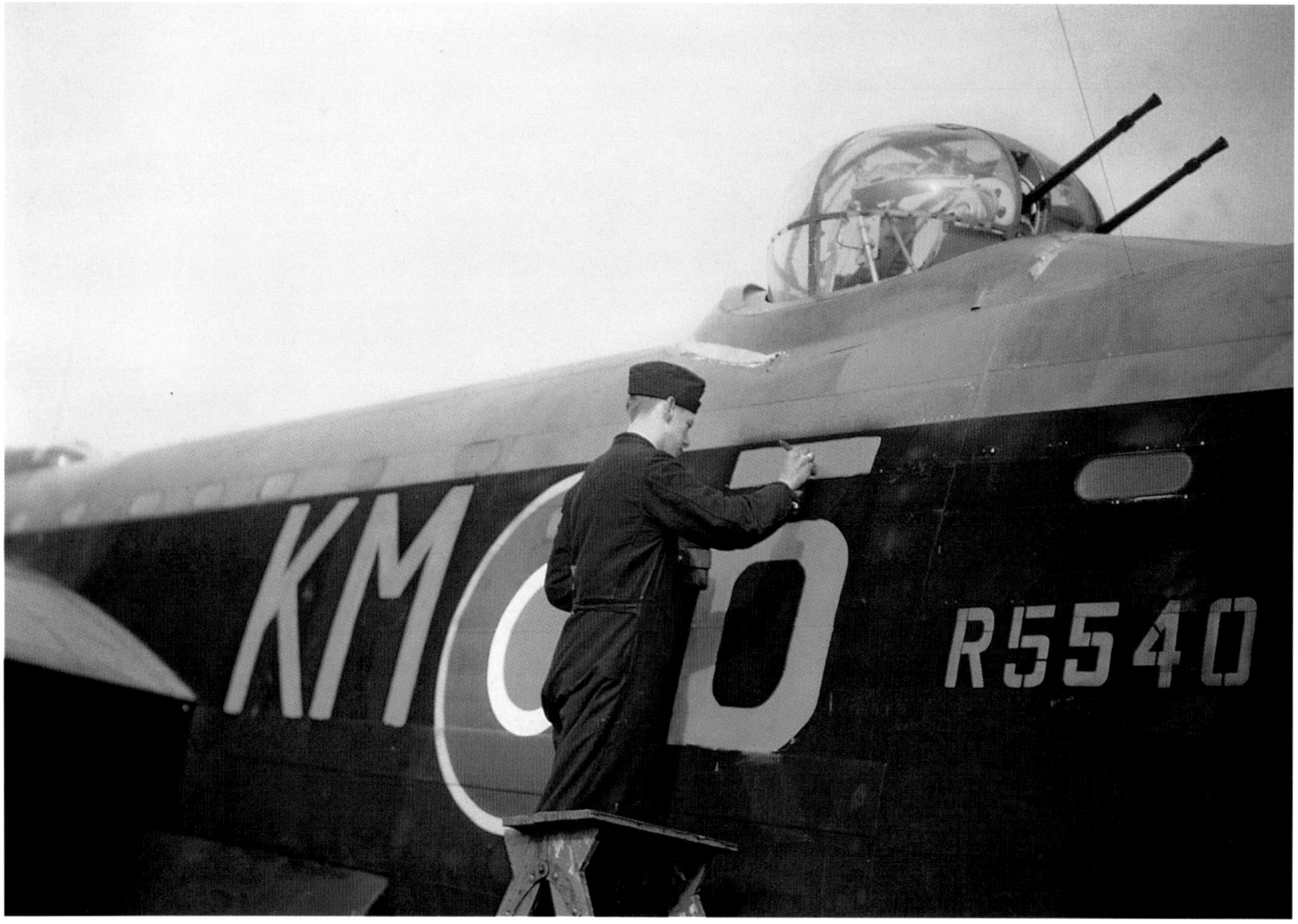
KM
R5540

THE GERMAN NIGHT FIGHTER AND THE RAF NIGHT BOMBER REPORT

RAF AND GAF AIRCREW COMPARE NOTES A.D.I (K) REPORT NO. 337/1945

All three pilots (German) considered that the bomber's tracer or gun flash did not give away the bomber's position, and SCHNAUFER remembers only two or three cases when bombers have been shot down due to this. He said he rarely saw the bomber's tracer and therefore, it did not worry him unduly while aiming. He was of the opinion that the bomber gunners did not fire nearly enough or soon enough. He was convinced that had the bombers fired more and used much brighter tracer, many of the pilots, the less experienced especially, would not have attacked. Asked if bright tracer fired from the fighter interfered with his aiming he said it did, and appreciated that it would be the same for the bomber gunner. SCHNAUFER had been surprised twice by fire from a bomber which he and not seen, and in each case the fire was accurate; these bombers he did not attack.

SECRET AT

ABOVE: A RIGGER TOUCHES UP THE CODE LETTERS KM-O JUST BELOW THE FN 50 TOP TURRET OF LANCASTER R5540.
RIGHT TOP: THE GUNNER'S VIEW FROM THE TOP TURRET. RIGHT BOTTOM: THE GUNNER'S VIEW FROM THE FN 20 TAIL TURRET.
BATTLE OF BRITAIN MEMORIAL FLIGHT.

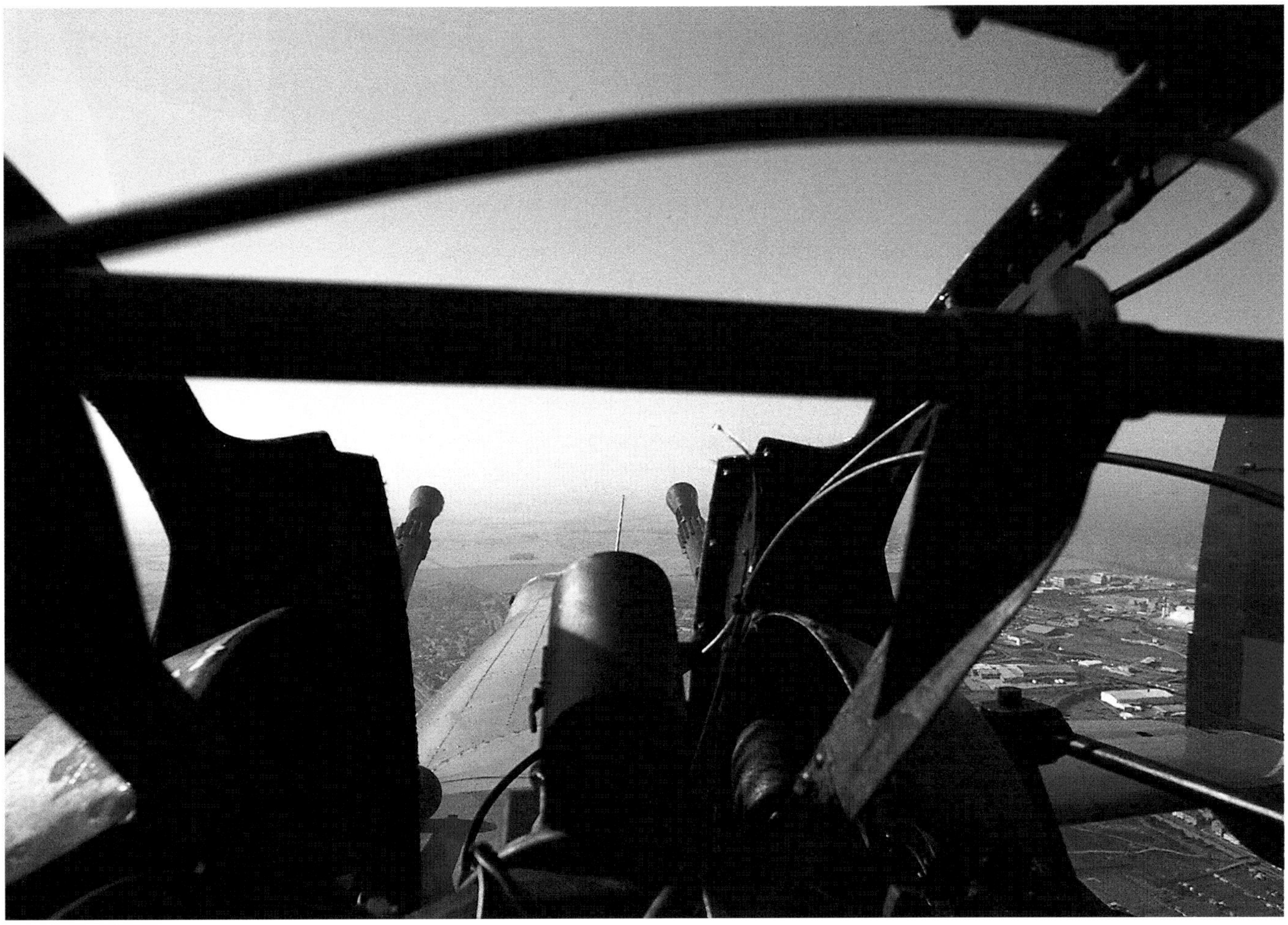

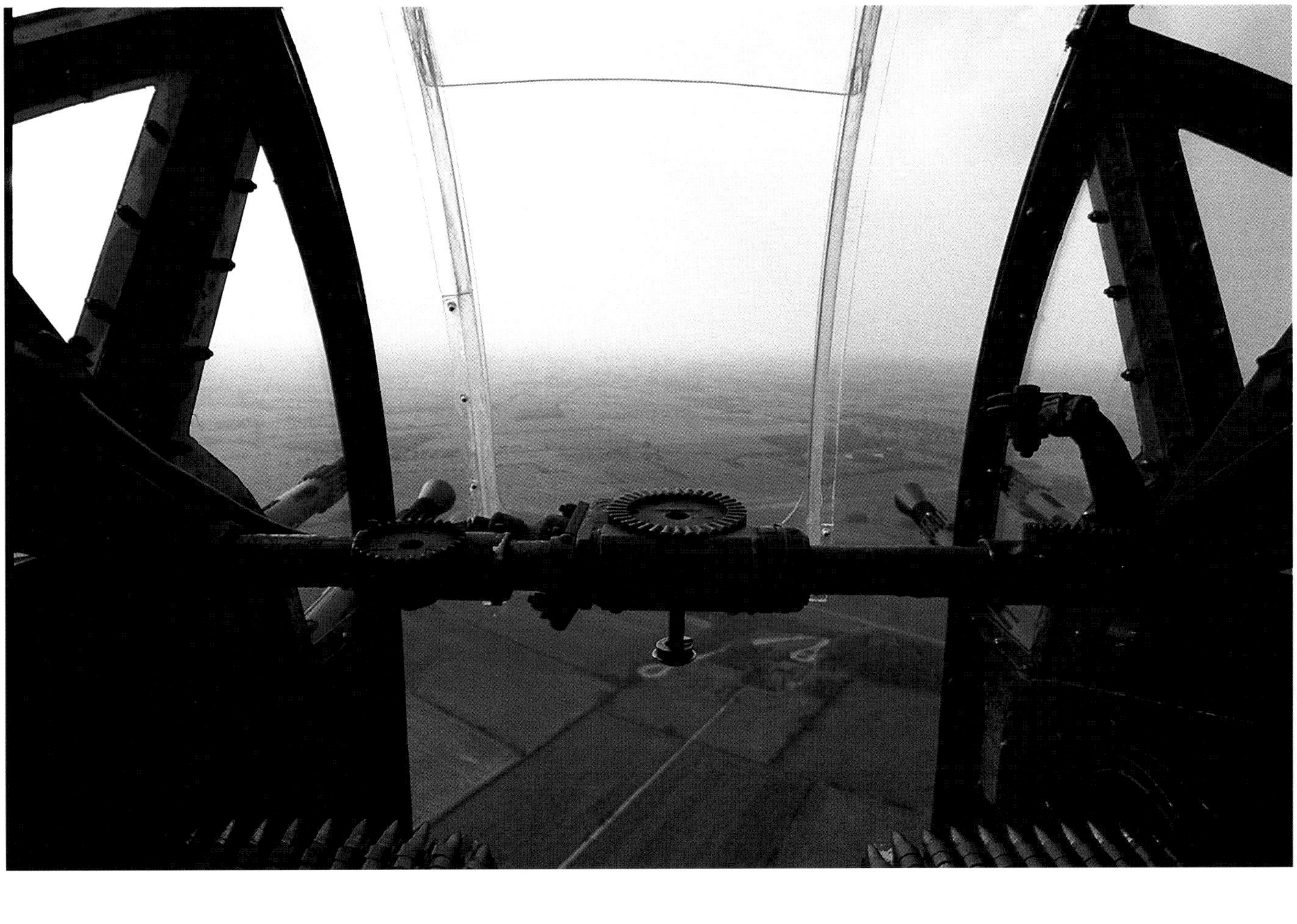

1. CUPOLA
2. OPEN SIGHT PANEL (RAISED)
3. CUPOLA BACK ASSEMBLY
4. INSTRUMENT PANEL
5. DIMMER SWITCH
6. WARNING LIGHT UNIT
7. FLOODLIGHT
8. GUN SIGHT
9. SIGHT RADIUS ARM
10. SIGHT ARM
11. GIMBAL RING
12. SIGHT LINKAGE
13. REAR SLING ATTACHMENT
14. INNER SIDE FRAME
15. BROWNING ·303 IN. GUNS (R.H.)
16. AMMUNITION BELT GUIDE
17. SERVO FEED MECHANISM
    PLATENS ★
18. SERVO FEED MECHANISM ★
19. SERVO FEED MECHANISM
    CLUTCH UNIT ★
20. GUN SLOT SEAL
21. FRONT SLING ATTACHMENT
22. GUN CRADLE
23. GUN FRONT ATTACHMENTS
24. OXYGEN FLEXIBLE PIPE
25. ASSISTER LEVER
26. MASTER VALVE HAND LEVER
27. FIRING TRIGGER
28. CONTROL HANDLE
29. FLEXIBLE PIPES TO REAR SEAR
    RELEASE UNITS ♦
30. REAR SEAR RELEASE UNIT ♦
31. OUTER SIDE FRAME
32. GUN REAR ATTACHMENT
33. FIXED GUN BRACKET
34. SERVO CONTROL VALVE ♦
35. SERVO FEED MECHANISM
    HYDRAULIC MOTOR FILTER ♦
36. CONTROL COLUMN
37. ELEVATION VALVE PUSH-ROD
38. BOWDEN CABLES
39. TOOL BAG
40. CUPOLA DOOR (R.H.) ■
41. CUPOLA DOOR CATCH ■
42. GUN ELEVATION RAM (LONG) ♦
43. UPPER ELBOW JOINT ♦
44. FLEXIBLE CONNECTING PIPE ♦
45. CARTRIDGE TRAYS
46. CARTRIDGE EJECTION CHUTE
47. GUN ELEVATION RAM (SHORT)♦
48. FIRING CONTROL VALVE ♦
49. HYDRAULIC ACCUMULATOR ♦
50. OXYGEN FLOWMETER
51. HYDRAULIC MOTOR ●♦
52. HAND ROTATION GEAR
53. FILTER ♦
54. VALVE BOX ♦
55. RAM BEARING TUBE
56. ROTATION LOCK
57. HYDRAULIC PIPES ♦
58. IMPEDANCE MATCHING UNIT
    PLUG
59. IMPEDANCE MATCHING UNIT

60. STOWAGE BAG
61. SAFETY BELT
62. SEAT
63. TURRET FIXED RING
64. TURRET ROTATING RING
65. VERTICAL ROLLER
66. INTERCOMMUNICATION SOCKET
67. LOWER ELBOW JOINT ♦
68. ACCOMMODATION PLATE
69. GEAR RING
70. HORIZONTAL ROLLER
71. FLEXIBLE PIPE TO EXTERNAL
    ROTATION VALVE ♦
72. TURRET DRUM
73. FIXED FLOOR PLATE

74. FLOOR PLATE VERTICAL ROLLER
75. ROTATING SERVICE JOINT
76. AMMUNITION FLOOR GUIDE ★
77. FLOOR PLATE HORIZONTAL
    ROLLER
78. OXYGEN SUPPLY PIPE
79. ELECTRICAL LEADS
80. MAIN OIL RETURN PIPE ♦
81. MAIN OIL SUPPLY PIPE ♦

● SEE A.D. 2252 - HYDRAULIC
            MOTOR

■ SEE A.D. 2499 - OPENING AND
    CLOSING OF ACCESS DOORS

♦ SEE A.D. 2559 SHEETS I AND
    2 - HYDRAULIC SYSTEM

★ SEE A.D. 2560 SHEETS I AND
    2 - SERVO FEED MECHANISM

TAIL GUN TURRET
(TYPE F.N. 120)

(FOR FURTHER INFORMATION
    SEE A.P. 1659A
    VOL. I AND VOL. II)

AIR DIAGRAM 2558		
SHEET I	No. OF SHEETS I	
JUNE 1945		

PREPARED BY
MINISTRY OF
AIRCRAFT PRODUCTION
FOR
PROMULGATION BY
AIR
MINISTRY

ABOVE: AN RAF AIR DIAGRAM CUTAWAY DRAWING OF THE FN 120 TAIL TURRET. THE FN 120 WAS A MODIFIED VERSION OF THE FN 20.
RIGHT: THE FN 20 WAS THE STANDARD TAIL DEFENSE OF THE AVRO LANCASTER, VICKERS WELLINGTON AND SHORT STIRLING.
MOST TURRETS, WHEN OPERATING AT HIGH ALTITUDE, SUFFERED FROM FROST GLAZING ON THE FRONT PERSPEX.
MOST GUNNERS REMOVED THE FRONT PANEL. THIS MEANT A COLDER TURRET, BUT THE CLEAR VIEW WAS OFTEN THE
DIFFERENCE BETWEEN SEEING THE ENEMY FIRST OR NOT SEEING HIM AT ALL.
CANADIAN WARPLANE HERITAGE MUSEUM.

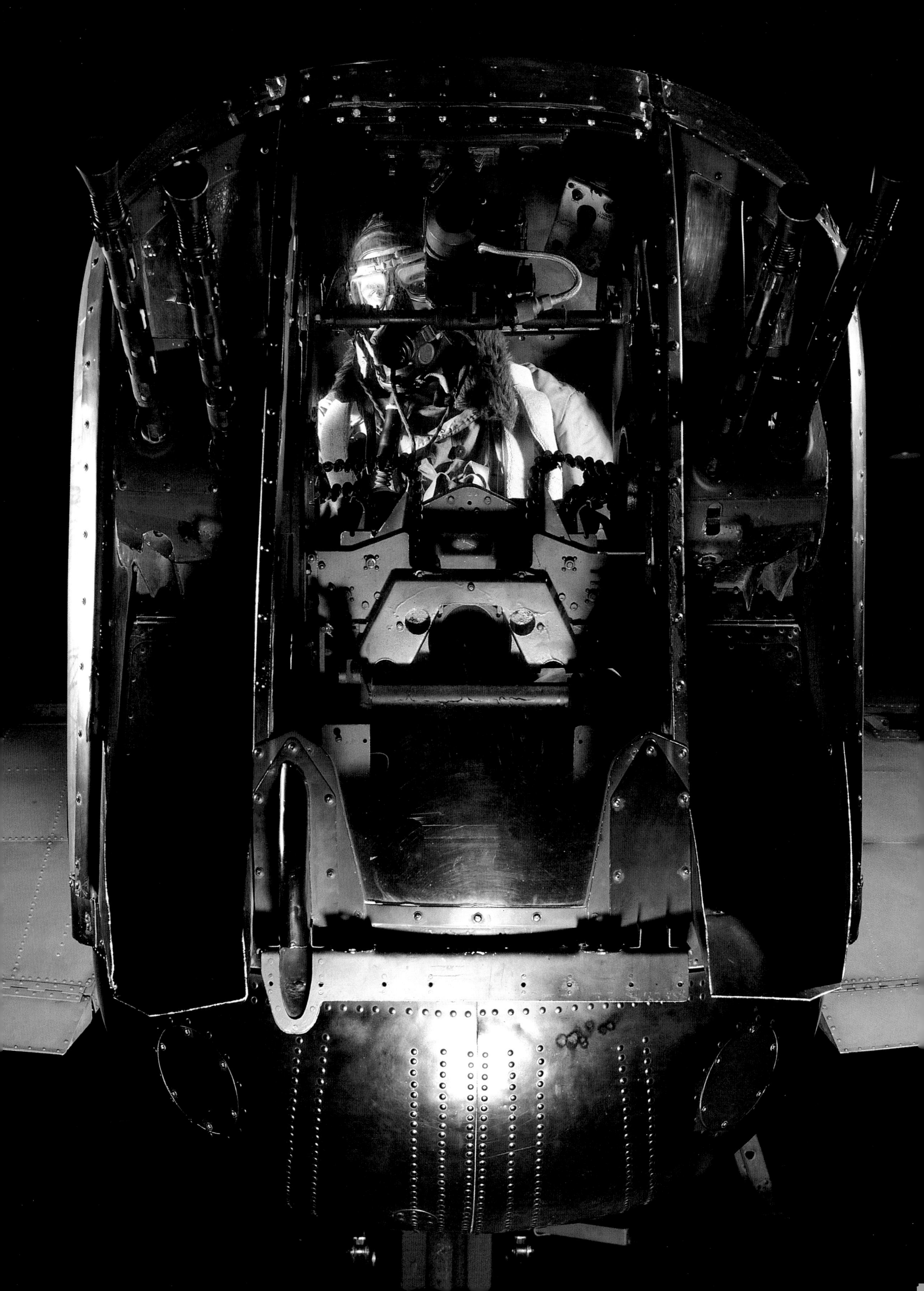

Handley Page Halifax

Never as popular as the Lancaster, the Halifax did offer its crews a better rate of survival. Of Halifax crews shot down, 29 percent survived, compared to 17 percent for the Stirling and only 11 percent for the Lancaster.

Constantly overshadowed by the Avro Lancaster, the Handley Page Halifax was a major contributor to the bombing campaign against Germany. Never able to lift the same bomb load as the Lancaster, or fly as far, the Halifax was far better than its critics credited it. Although its loss rate was higher than the Lancaster, its crew survival rate was much better. Only 11 percent of Lancaster crew survived, but 29 percent of Halifax crewman shot down survived. The Halifax's fuselage was more capacious, which made movement easier, especially when it came time to bail out.

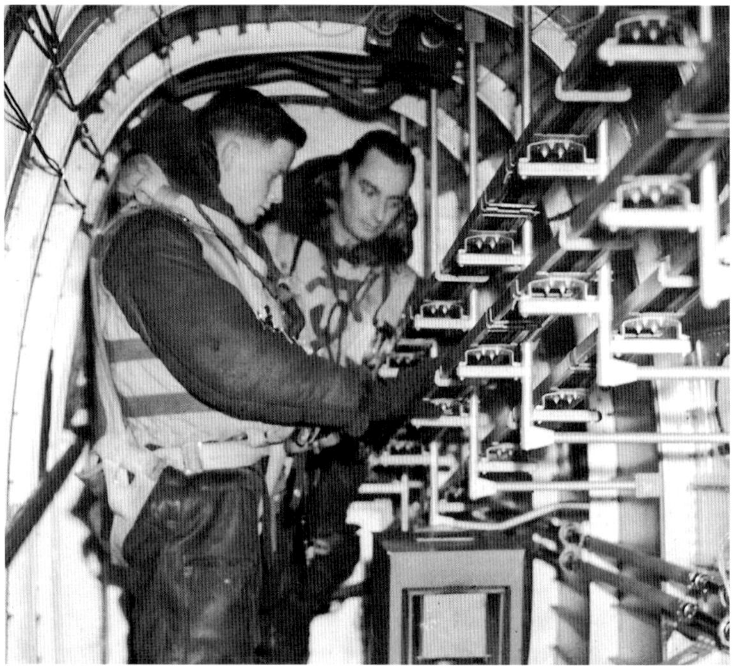

When the Halifax B Mk I first entered service, it was equipped with Boulton Paul powered turrets in the nose and tail, each equipped with .303-caliber Browning machine guns (two in the nose, four in the tail), along with two hand-held Vickers .303-caliber VGO machine guns that could be aimed through the beam hatches. The nose turret was the Boulton Paul Type C Mk I. Although it was not used a great deal, it did give some protection against head-on attacks during daylight missions. The rear turret was the Boulton Paul Type E tail turret, armed with four .303-caliber Brownings. The E Type turret was very popular with the gunners and was fitted to all Halifax aircraft. It was not until the last months of the war that the Halifax was equipped with the more powerful (two .50-caliber machine guns) Type D tail turret. The Halifax Mk I was followed by the B Mk II armed with a Boulton Paul Type C Mk II mid-upper turret. This turret reduced speed by 6 mph and was soon followed by the more streamlined Type A Mk VIII four-gunned turret.

The most successful Halifax was the Mk III. This aircraft removed the nose turret and replaced it with a streamlined, clear perspex nose and mounted a single free-mounted Vickers VGO "scare gun." Why they did not replace the nose turret in the Lancaster is a mystery. In the night war above Germany, it was useless. The new Mk III was faster and had a higher ceiling with a greater range.

Many Halifaxes, especially in 6 Group (Canadian), installed the Preston Green under defense turret. It was not a powered turret by definition, but a single .50-caliber machine gun mounted in the floor of the aircraft. If more aircraft had been fitted with the Preston Green, the toll exacted by Luftwaffe night fighters equipped with the deadly *schräge Musik* upward-firing cannon may have been lessened. Instead, when H2S radar production increased, the turrets were taken out. In the end, the Halifax equipped no fewer than thirty-five squadrons in Europe, with four more in the Middle East.

Above: Gunners check the ammunition feeds to their guns before take-off.

Right: Two Halifax Mk IIs in formation. The Mk II was armed with three Boulton Paul powered turrets. Later versions dispensed with the front turret and replaced the rather large, bulbous Type C Mk II upper turret with the more streamlined, four-gunned Type A turret.

Inset: A rigger cleans the cockpit perspex while an armorer rods the barrels of Boulton Paul Type C Mk I nose turret.

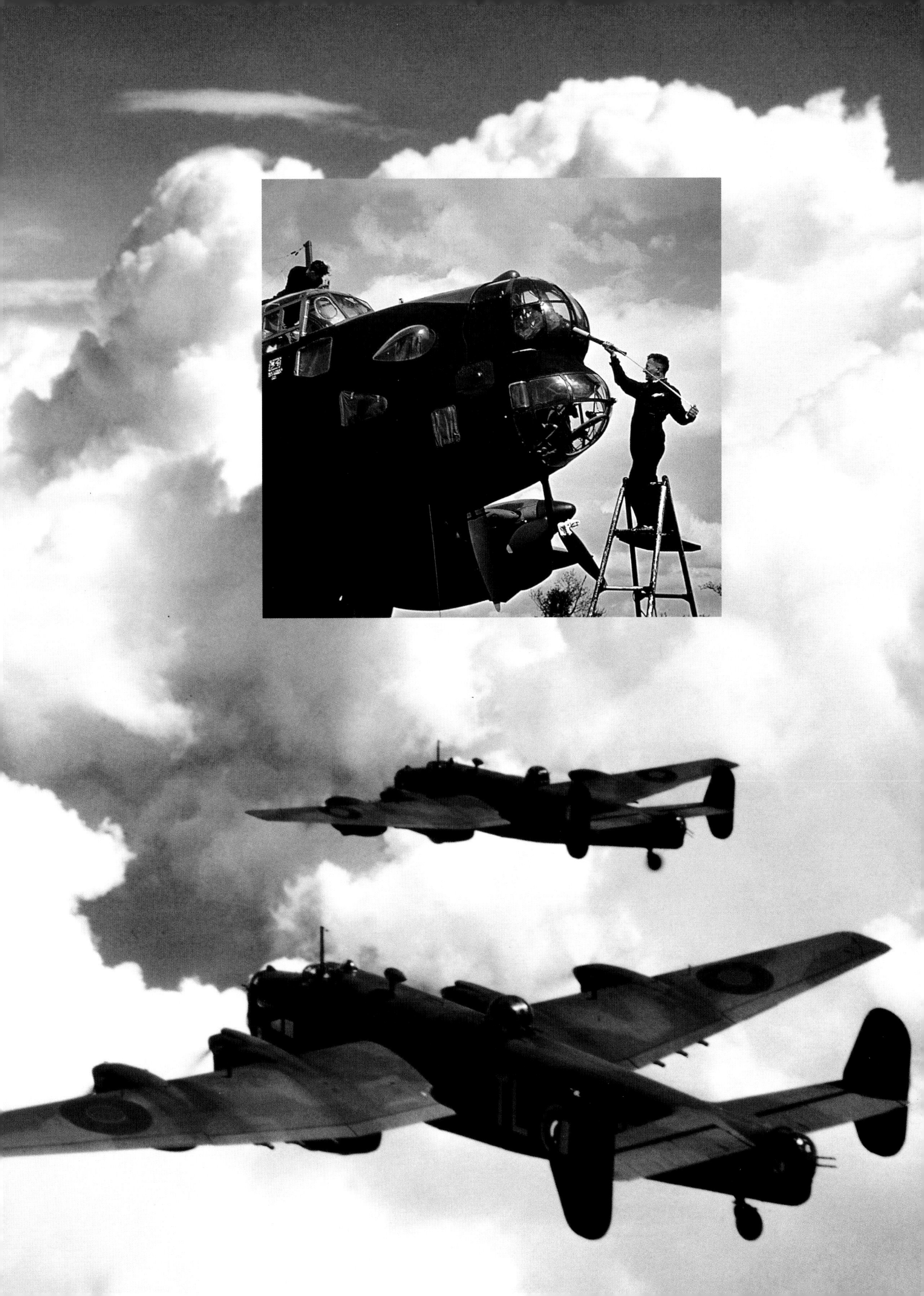

"From the Whitley, I went to Dishforth for conversion onto Halifax IIs. The rear turret in the Halifax was a Boulton Paul with four Brownings mounted on their sides, instead of up and down like in the Frazer Nash turret in the Lancaster. I found the Boulton Paul turret had too many bulgy perspex panels, which caused distortions. In the Frazer Nash turret, we cut the front perspex out. This prevented the condensation from our breath from freezing on the perspex at forty below.

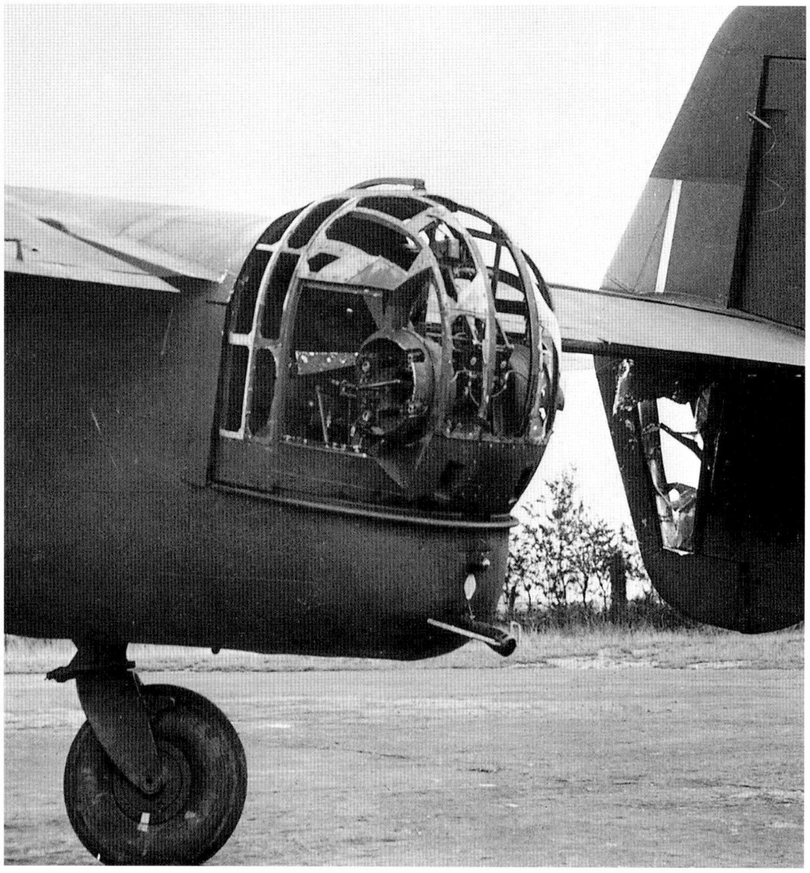

"As part of our night-vision training, we were taught to look out the bottom of our eyes. When you look up you also look down at the same time. I still play that game when I'm driving. It's amazing what you actually see."

FLYING OFFICER BILL COLE,
HALIFAX AND LANCASTER TAIL TURRETS,
428 SQUADRON, 6 GROUP, RCAF

"One of the things you had to always deal with on long trips was staying awake. That was very difficult. The longest trip I did was about eight hours, and that's a long time to be looking at nothing. It was hard on the eyes; you had to practically hold them open.

"My first trip was a daylight trip to Aerodrome in Holland. I figured, boy, this is sure going to be an easy life. There wasn't a cloud in the sky, no flak, not a fighter to be seen. I figured, this is great, but the next night we went to Kiel and that was a different story. We almost hit a Junker Ju 88 or Bf 110 head-on over the target. I say he missed by about ten or fifteen feet. They just close so fast you never saw them — bang, he's gone! I liked the Halifax because it was so much easier to get out of. If you had a problem, there was so much more room. If a Lanc and a Hali got shot down, 1.5 guys got out of the Lancaster and 2.5 got out of the Halifax.

"One of the worst times I had — I'd done about twenty-five trips — so I'm flying with this crew as a spare and we are going to Magdaberg. This was their second trip. So we are going across the North Sea at about 15,000 feet when I spotted smoke coming from the port engine. The pilot looks out and says, 'No, that's just mist spraying over the wing.' A bit later, I said, 'I'm sure that's smoke coming from the port inner.' The Engineer comes on: 'The revs are okay, the oil pressure is good and everything else.' Five minutes later I said, 'Well, that engine is starting spark and there's flame coming out!' 'Feather the port inner.' Now the plane wouldn't climb, and we were approaching Heligoland. So we had to turn back. Flying back, were we now even lower, but you couldn't cross the English coast under a certain height. So the pilot's trying to climb, but he wasn't having too much success. So he told the bomb aimer to drop the bombs and make sure to make them safe. The bomb aimer said, 'You don't have to tell me, I know my job.' A moment later, he said, 'Bombs gone.' All I heard was whomp, whomp! Those bombs blew us up about a thousand feet and by then we were high enough cross the English coast. I was glad I didn't have to fly with them anymore."

JOE DAVIS, WIRELESS AIR GUNNER,
429 SQUADRON, 6 GROUP, RCAF

ABOVE: GERMAN NIGHT FIGHTERS COMPLETELY OUTGUNNED BRITAIN'S HEAVY BOMBERS. THIS TURRET SUFFERED HEAVY DAMAGE AFTER AN ENCOUNTER WITH A GERMAN NIGHT FIGHTER OVER BERLIN.

RIGHT: THE BOULTON PAUL TYPE E TAIL TURRET PROVIDED REAR DEFENSE FOR THE HALIFAX AND LIBERATOR II. IT WAS ONE OF THE MOST SUCCESSFUL TURRETS EVER. OVER 8,000 TYPE E TURRETS WERE PRODUCED DURING THE WAR.
RCAF MUSEUM, TRENTON.

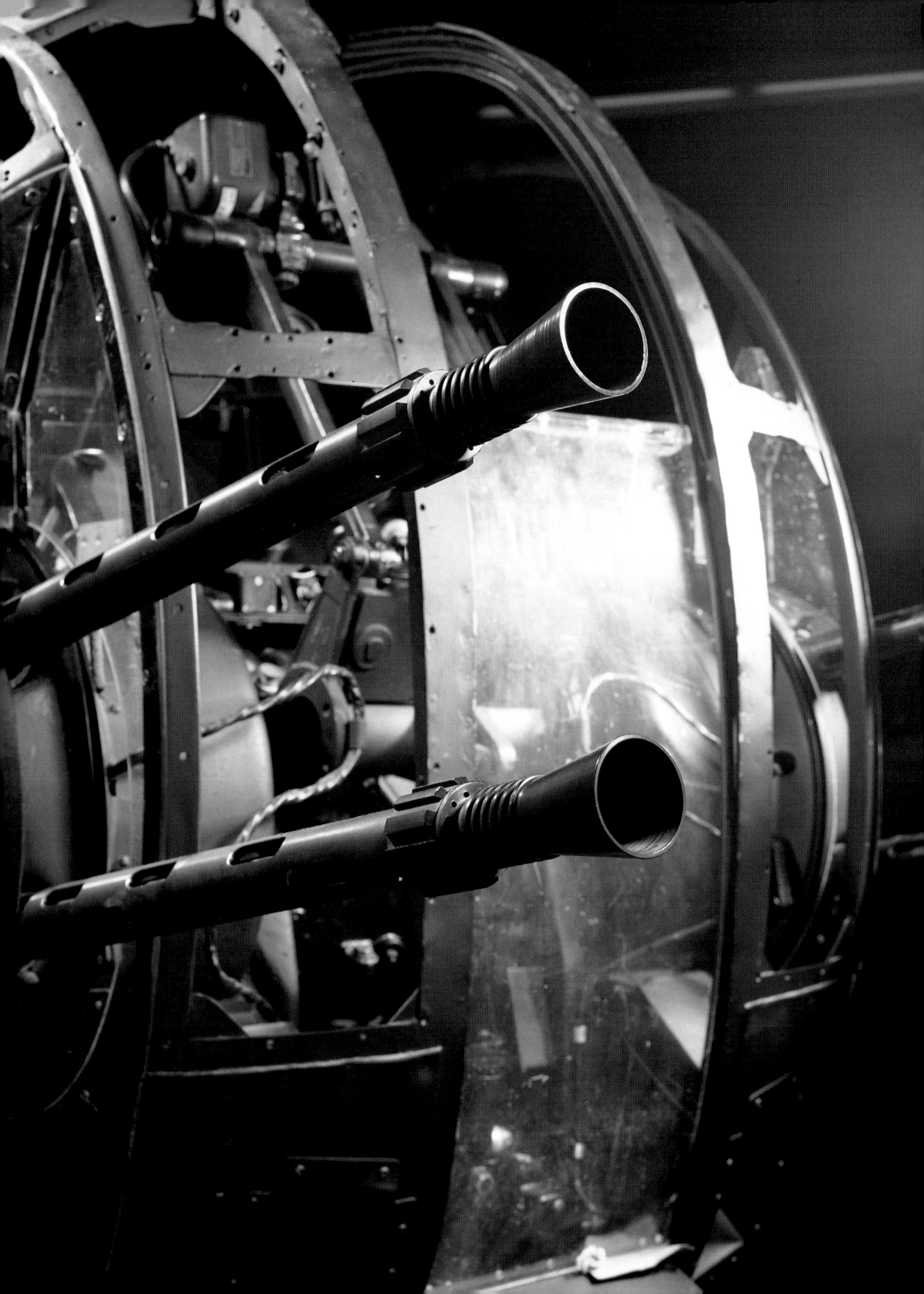

"When I finally became the permanent rear gunner in a crew, I was very content to remain a rear gunner. Escaping in an emergency was easier from this position, and I felt it was a good trade-off for the 'nicer' view in the mid-upper. It was certainly warmer and roomier than the experiences in the mid-upper, which was extremely cold and cramped.

"There were very few gunners who actually fired their guns in anger. There was a rumor going around one time about some chaps that had several kills. It was claimed that they actually kept a cigarette glowing or a flashlight to attract a night fighter. Whether it was true or not, nobody knows for certain — only the chap himself, and I doubt he would ever have admitted to it at the time. Perhaps he may today."

FLYING OFFICER DOUG SAMPLE, HALIFAX,
415 SQUADRON, 6 GROUP, RCAF

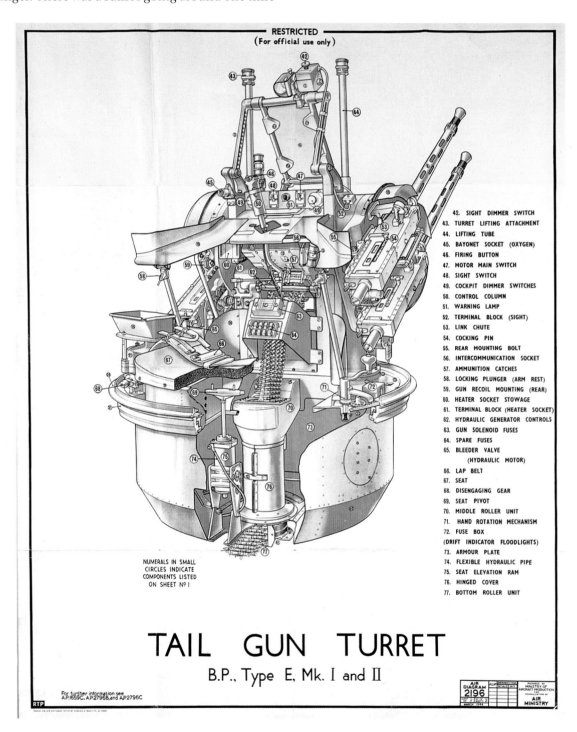

42. SIGHT DIMMER SWITCH
43. TURRET LIFTING ATTACHMENT
44. LIFTING TUBE
45. BAYONET SOCKET (OXYGEN)
46. FIRING BUTTON
47. MOTOR MAIN SWITCH
48. SIGHT SWITCH
49. COCKPIT DIMMER SWITCHES
50. CONTROL COLUMN
51. WARNING LAMP
52. TERMINAL BLOCK (SIGHT)
53. LINK CHUTE
54. COCKING PIN
55. REAR MOUNTING BOLT
56. INTERCOMMUNICATION SOCKET
57. AMMUNITION CATCHES
58. LOCKING PLUNGER (ARM REST)
59. GUN RECOIL MOUNTING (REAR)
60. HEATER SOCKET STOWAGE
61. TERMINAL BLOCK (HEATER SOCKET)
62. HYDRAULIC GENERATOR CONTROLS
63. GUN SOLENOID FUSES
64. SPARE FUSES
65. BLEEDER VALVE
       (HYDRAULIC MOTOR)
66. LAP BELT
67. SEAT
68. DISENGAGING GEAR
69. SEAT PIVOT
70. MIDDLE ROLLER UNIT
71. HAND ROTATION MECHANISM
72. FUSE BOX
(DRIFT INDICATOR FLOODLIGHTS)
73. ARMOUR PLATE
74. FLEXIBLE HYDRAULIC PIPE
75. SEAT ELEVATION RAM
76. HINGED COVER
77. BOTTOM ROLLER UNIT

NUMERALS IN SMALL
CIRCLES INDICATE
COMPONENTS LISTED
ON SHEET Nº 1

TAIL GUN TURRET
B.P., Type E, Mk. I and II

For further information see
A.P.1659C, A.P.2796B, and A.P.2796C

AIR DIAGRAM 2196
MARCH 1944
PREPARED BY MINISTRY OF AIRCRAFT PRODUCTION
AIR MINISTRY

ABOVE: RAF AIR DIAGRAM OF THE BOULTON PAUL TAIL GUN TURRET.

RIGHT: THE BEAUTIFULLY RESTORED ALL-BLACK INTERIOR OF A BOULTON PAUL TYPE E TAIL TURRET. THIS TURRET IS PART OF THE RCAF MUSEUM'S RESTORATION OF A HALIFAX MK C. MK VII. THIS AIRCRAFT HAD THE TASK OF DROPPING SPECIAL AGENTS AND OR SUPPLIES BY PARACHUTE INTO ENEMY TERRITORY.

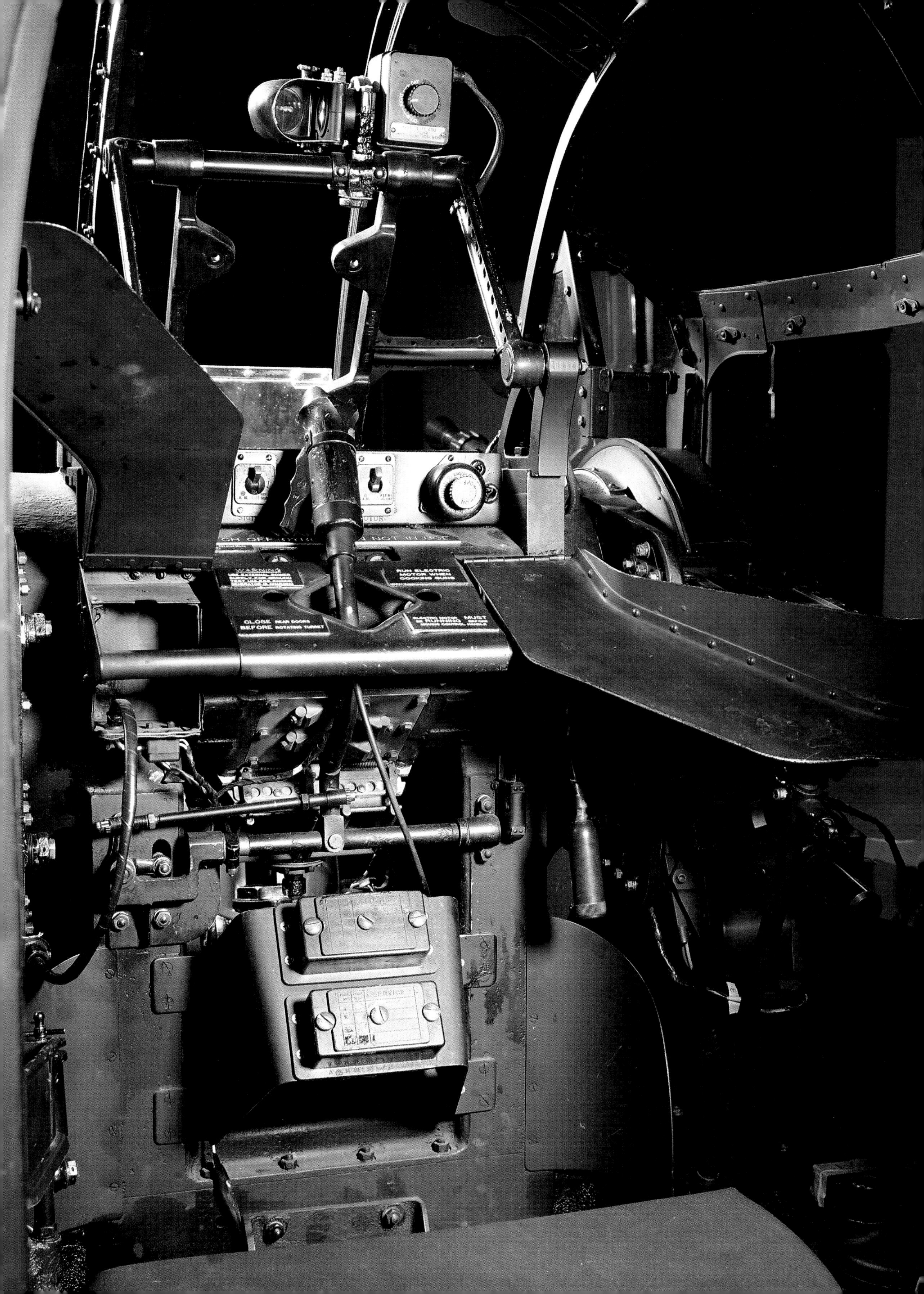
WARNING
RUN ELECTRIC MOTOR WHEN COCKING GUNS
CLOSE REAR DOORS BEFORE ROTATING TURRET
MUST BE RUNNING

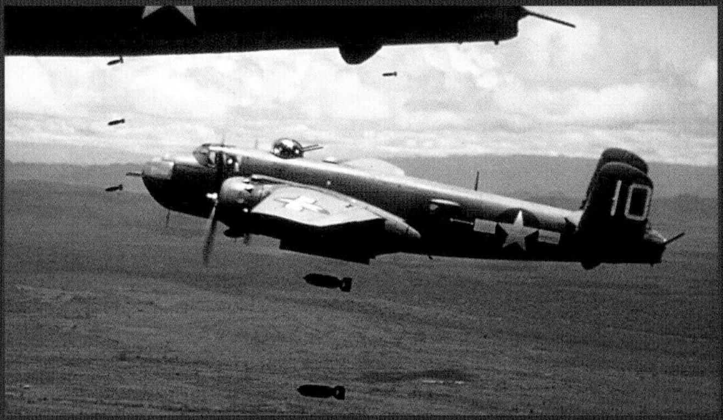
10

North American B-25 Mitchell

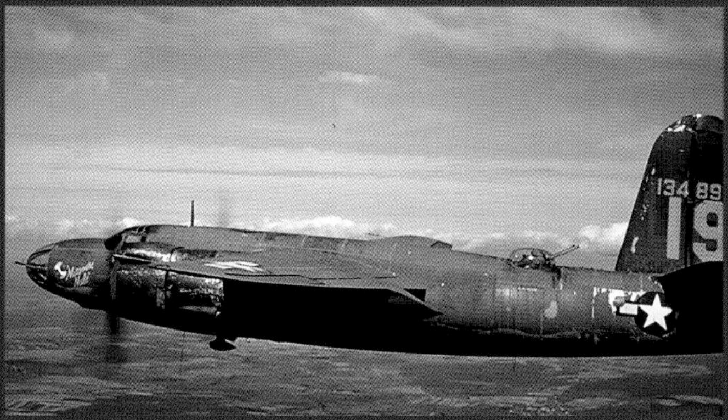
134 89

Martin B-26 Marauder

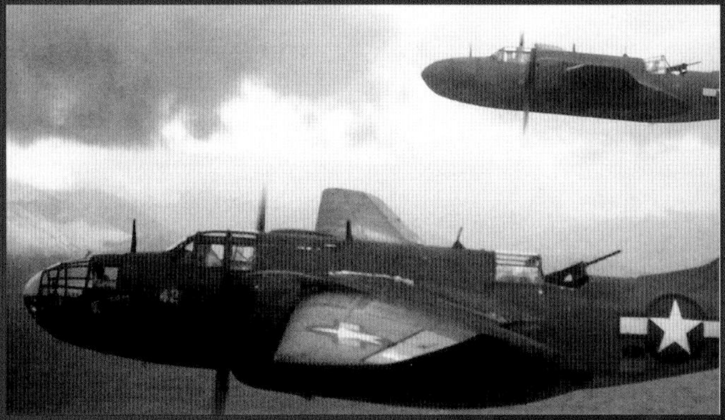

Douglas A-20 Havoc

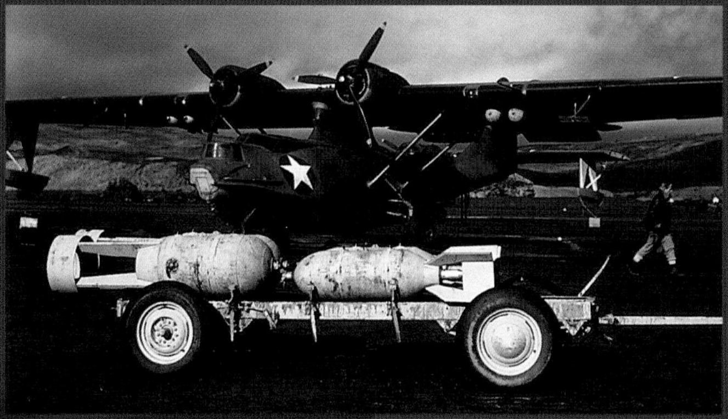

Consolidated PBY Catalina

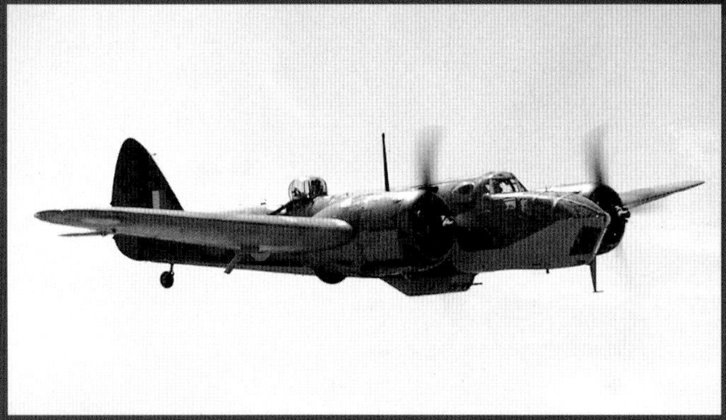

Bristol Blenheim

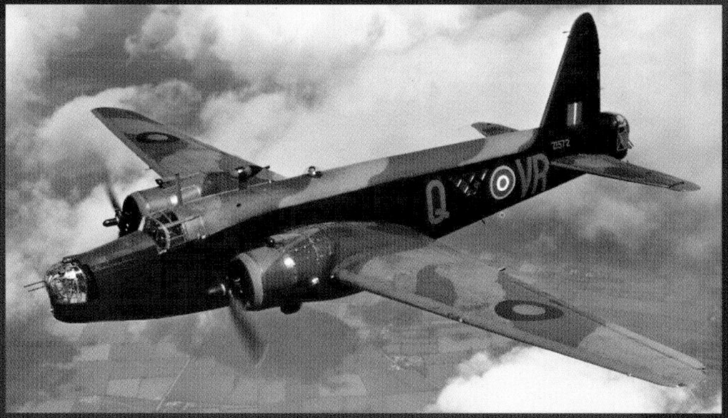
Q
VR

Vickers Wellington

Twin Engine

✠ Messerschmitt Bf 110

✠ Junkers Ju 88

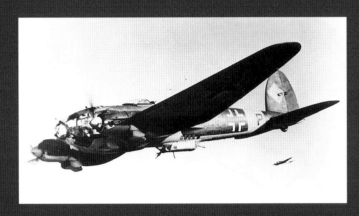

✠ Heinkel He 111

✠ Messerschmitt Me 410

Savoia-Marchetti S.M. 79
Sparviero

NORTH AMERICAN B-25
MITCHELL

THE B-25 MITCHELL WAS ONE OF THE MOST HEAVILY ARMED BOMBERS OF THE SECOND WORLD WAR.
IT SAW ACTION ON EVERY MAJOR BATTLEFRONT AND WAS ARGUABLY THE BEST
MEDIUM BOMBER OF THE WAR.

The medium bombers of World War II were not subject to the heavy fighter attacks suffered by their heavy-bomber brethren. While the B-17, B-24 and B-29 were pummeled by the best the Luftwaffe and Japanese fighter force had to offer, the medium bombers such as the B-25, B-26, A-20, Ventura and Blenheim usually operated with fighter escort. But even with their escort, they were still armed with powered turrets and hand-held machine guns — and one of the most heavily armed was the North American B-25 Mitchell. Early versions of the B-25 were lightly armed with a single .30-caliber machine gun in the nose, two .50-caliber guns in the Bendix top turret, and two in the retractable Bendix ventral turret. The Bendix ventral turret was not a success. Fitted to the B-24 and B-25, it had an inadequate field of view, so the gunner found it extremely difficult to track a moving target. Waist guns (with limited fields of fire) were fitted to cover the bottom of the aircraft. One of the most heavily armed B-25s was the H model. These carried eighteen .50-caliber machine guns, five more than the B-17, the so-called Flying Fortress. It most be noted that twelve of the guns fitted were not for defense against fighter attack, but for anti-shipping strikes and ground attack.

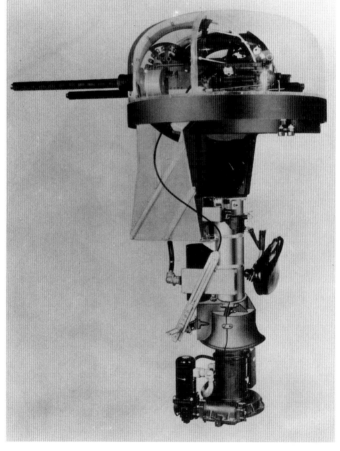

The J model B-25 was produced in greater numbers than any other version of the Mitchell. This aircraft was fitted with no fewer than twelve .50-caliber machine guns. The Bendix top turret had two guns and was located just behind the pilot compartment. This gave the gunner a good view and even better field of fire. A Bell power turret provided the sting in the tail. The waist position carried two .50-caliber guns, and in the nose there was one flexible gun and one fixed. To supplement the nose armament, there were four blister guns mounted on the sides of the fuselage just below the pilot and co-pilot.

The B-25 Mitchell saw combat in every theater, building up an unrivaled reputation, and was arguably the best medium bomber of the war.

ABOVE: THE BENDIX UPPER TURRET WAS USED ONLY IN THE MITCHELL.
IT CARRIED 400 ROUNDS PER GUN AND WAS ELECTRICALLY POWERED.
RIGHT: WARTIME PROPAGANDA EXTOLLED THE EFFECTIVENESS OF TURRET-ARMED BOMBERS.
THIS AD FOR NORTH AMERICAN AVIATION MAKES IT LOOK AS IF THE B-25 WAS IMMUNE TO FIGHTER ATTACK.

ZERO ÷ B-25 = 💥

Tojo's boy's are learning simple arithmetic from Uncle Sam's...

Typical is this flyer's report:

"Newspaper stories of American flyers knocking down Jap planes by the dozens, while losing only a few, are absolutely correct. I remember the gunner of my B-25... as soon as the Japs got within range of his .50 caliber machine gun, he began firing bursts to keep the Jap away. He protected our ship very well, and when there was a Jap pilot with guts enough to come in close,

the gunner made it so hot for him that the guy sheered off. This gunner got three Zeros.

"Another time, our B-25's and fighter escort were jumped by 40 or 50 Zeros. We got 23 for sure, probably 9 more, and lost one plane. The ratio of Jap losses to ours is really high!"

That's just a pilot's way of saying that Zero divided by B-25 equals zero—and always will!

The men and women of North American Aviation are delighted to have a hand in the education of the Sons of Heaven... proud to supply, as the lethal divisor in this simple

problem in American arithmetic, the B-25 Billy Mitchell bomber—"the old reliable," flyers call it.

North American Aviation planes, brilliantly engineered and soundly built, have had no small share in the outstanding achievements of American flyers on every front of this global war.

North American Aviation, Inc., designers and builders of the B-25 Mitchell bomber, the AT-6 Texan combat trainer and the P-51 Mustang fighter (A-36 fighter-bomber). Member of the Aircraft War Production Council, Inc.

North American Aviation *Sets the Pace!*

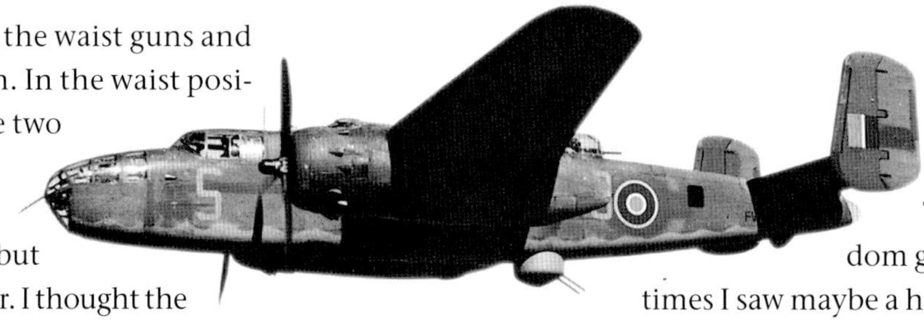

"I was assigned the waist guns and the tail position. In the waist position, there were two .50-caliber (12.7-mm) machine guns, but only one gunner. I thought the .50-caliber was great; it could demolish anything it hit. The waist gun position was very comfortable. The guns were staggered and the hatches were covered with Plexiglas. You could sit there, look out the window, watching the world go by. The guns were well organized, easy to swing around and had plenty of ammunition. It wasn't really cold. We operated from ground level up to 10 or 12,000 feet. We worried more about flak than fighters. When you're flying around 10,000 feet, you're a pretty good target for flak. One time we were hit by flak that took the top off the pilot's perspex. The wind then howled through the aircraft. The wind was so strong it managed to lift off a large chunk of the bomb-bay opening, sending it flying through the fuselage and hitting me in the back of the head. If I didn't have my flak helmet on, I probably would have been hurt. We got a lot of holes that time.

"I never had the chance to shoot at a German fighter. We always had fighter cover. The German fighters very seldom got down close to us. A few times I saw maybe a half a dozen or so. I was kind of disappointed that I never fired my guns. I always hoped that some enemy fighter would go roaring by and I'd get a crack at him. I saw one Me 109 that was close enough to almost touch. He was going past us, front to back, with this Spitfire right on his tail. They were so close I could see the expression on both their faces as they went by.

"The tail position was very tiny and uncomfortable, but with a great view. It was like being in your own cockpit.

"We were well equipped with machine guns. If you add the amount of firepower we had with eighteen aircraft, then put them in three boxes of six in close formation, that's something I wouldn't want to attack!"

JOE OUELLETTE, B-25 WIRELESS AIR GUNNER,
226TH SQUADRON,
2ND TACTICAL AIR FORCE

Section IV
Paragraph 22

RESTRICTED
AN 01-60GD-2

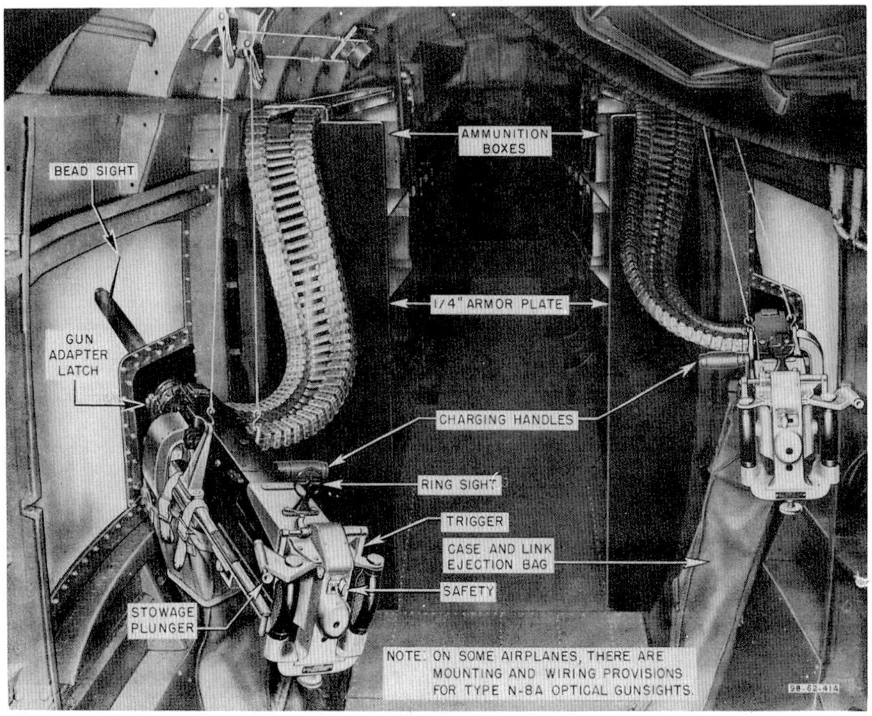

AMMUNITION BOXES
BEAD SIGHT
1/4" ARMOR PLATE
GUN ADAPTER LATCH
CHARGING HANDLES
RING SIGHT
TRIGGER
CASE AND LINK EJECTION BAG
SAFETY
STOWAGE PLUNGER
NOTE: ON SOME AIRPLANES, THERE ARE MOUNTING AND WIRING PROVISIONS FOR TYPE N-8A OPTICAL GUNSIGHTS.

Figure 459—Waist Gun Installation

TOP: AN RAF MITCHELL MK II WITH ITS BENDIX ELECTRICALLY OPERATED VENTRAL TURRET FULLY EXTENDED.
BOTTOM: THE B-25'S WAIST GUNNER'S POSITION.
RIGHT: EARLY VERSIONS OF THE MITCHELL WERE ARMED WITH A SINGLE .30-CALIBER NOSE GUN. LATER VERSIONS CARRIED ONE .50-CALIBER FLEXIBLE NOSE GUN, ONE FIXED GUN FIRING THROUGH THE FRONT PERSPEX, AND FOUR FIXED GUNS MOUNTED ON THE SIDE FUSELAGE, JUST BELOW THE COCKPIT.
USAF MUSEUM, DAYTON.

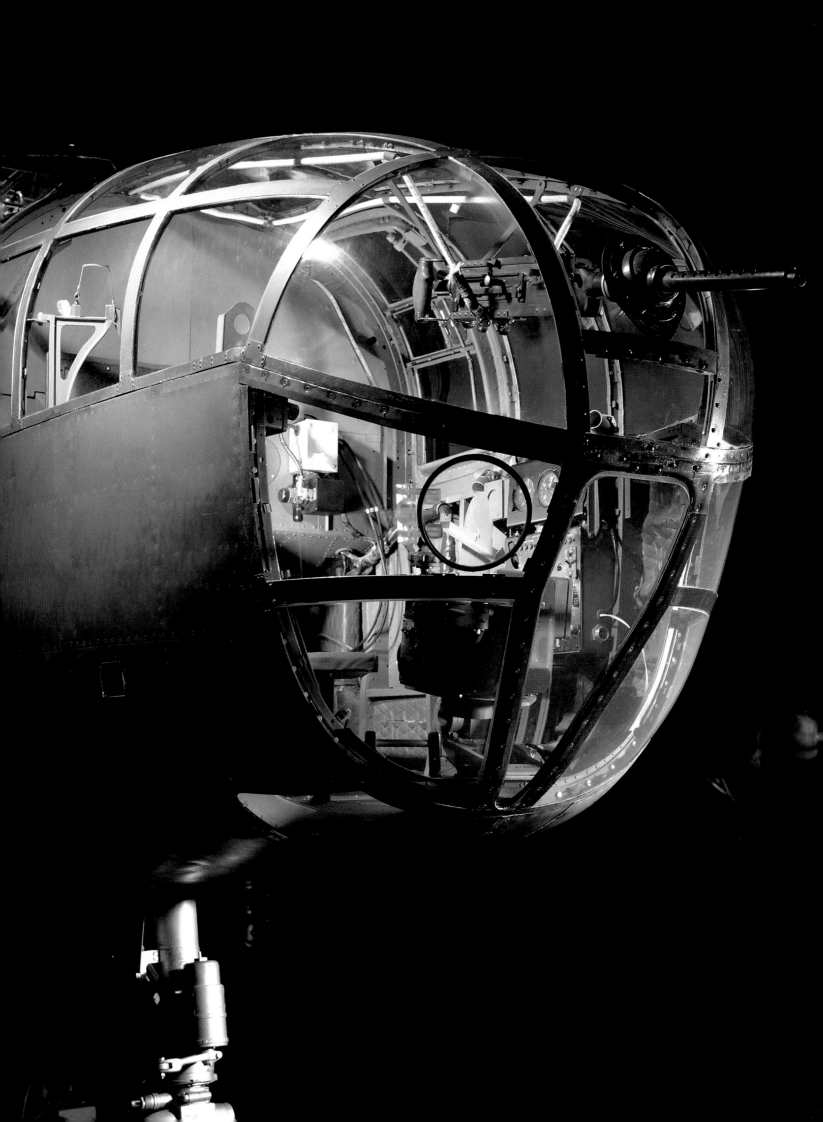

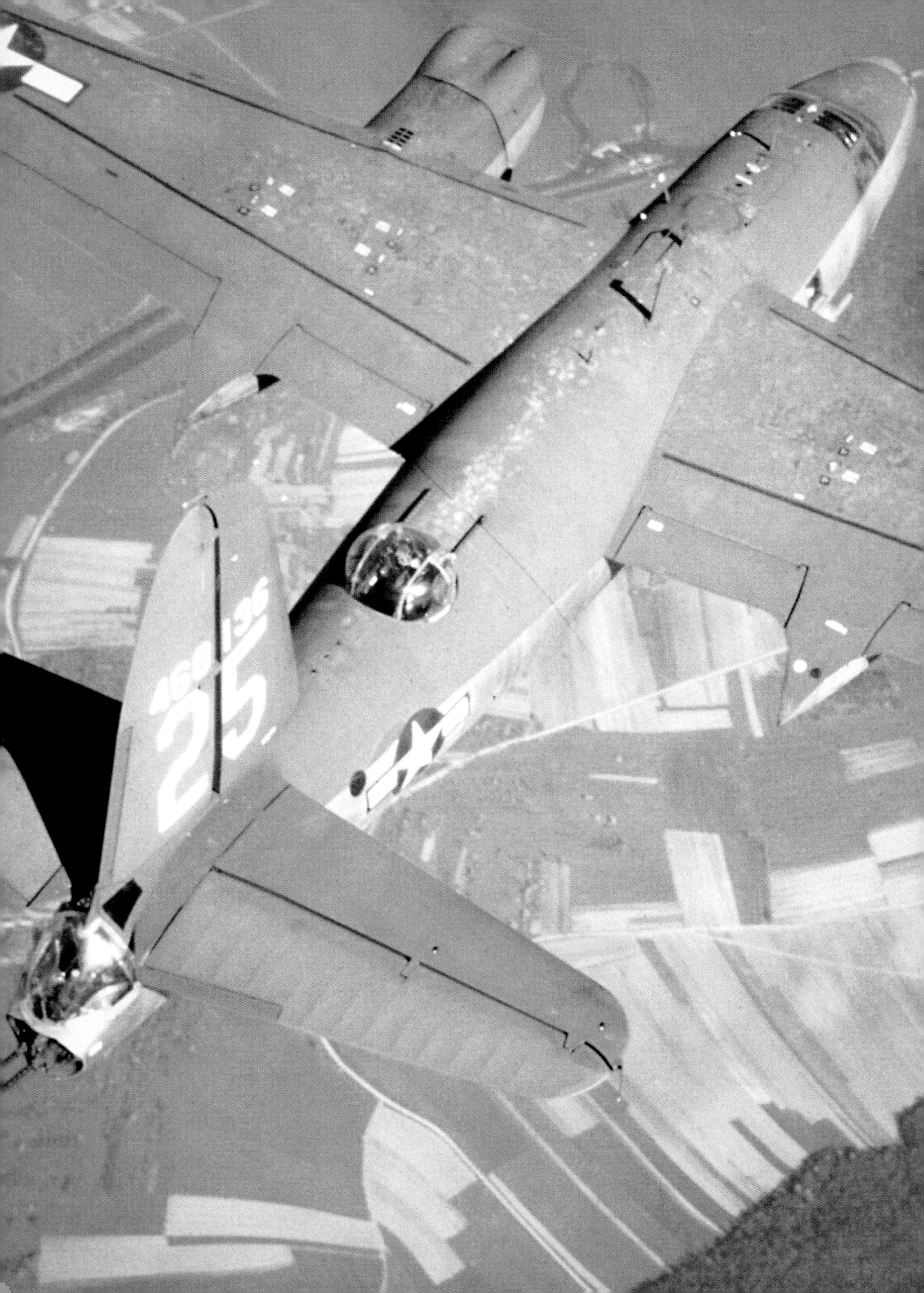
468136
25

MARTIN B-26
MARAUDER

ARMED WITH UP TO TWELVE .50-CALIBER MACHINE GUNS, THE MARAUDER COULD NOT ONLY DEFEND ITSELF, BUT FOR A BRIEF MOMENT TURNED THE TABLES AND BECAME A LONG-RANGE FIGHTER.

The Boeing B-17 was not the only Flying Fortress of the Second World War. When the B-26B-10 Marauder was introduced into service, it was armed with twelve .50-caliber Colt-Brownings, only one short of the B-17G! Mounted in the nose was one flexible .50-caliber with 270 rounds; this was augmented by a single fixed gun on the starboard side of the fuselage with 200 rounds. The forward armament was reinforced with two blister guns mounted on each side of the fuselage, with 200 to 250 rounds per gun (rpg). Completing the defensive armament were a low-drag electrically operated Martin 250CE dorsal turret with two .50-caliber machine guns (the Marauder was the first American bomber fitted with a power turret), two .50-caliber guns

in the rear waist hatches, and two machine guns in the tail. Later, the twin hand-held tail guns (early versions had one) were replaced by a power-operated Martin Bell turret with 400 rpg.

The Marauder was a fast, heavily armed medium bomber, but it also proved itself as a fighter. In the closing months of the Tunisian campaign, the Marauder's long range and heavy armament enabled it to intercept large formations of slow, unescorted, multi-engine Me 323 and Ju 52 German transports. The success of these operations caught the attention of the 8th Air Force. In July 1943, the heavy losses suffered by the American heavy bombers called for a long-range fighter. Army Air Force headquarters in the United States suggested that the B-26 could fulfill this role! The idea was not received with enthusiasm by 8th Air Force staff. Not only did the Marauder have an entirely different performance envelope from the B-17, but also, like all bombers, without fighter escort, it was extremely vulnerable. The idea was quickly dropped, and the B-26 went onto become one of the most effective medium bombers of the war.

LEFT: A 320TH BOMB GROUP MARAUDER OVER SOUTHERN ITALY. THE TOP TURRET GUNNER CAN BE SEEN MANNING HIS POSITION.
ABOVE: THE B-26B CARRIED TWO MANUALLY OPERATED .50-CALIBER MACHINE GUNS IN THE TAIL POSITION.

"My responsibility was to man the waist guns, which were the two in the back directly under the top turret. There was a gun on each side, and one of them you could pull up and use in the ventral tunnel hatch to fire out the bottom. I only went back to the waist guns when we were over the target, or shortly after we dropped bombs. That's when we were the most vulnerable. With the B-26, we bombed from low altitude — 12, 14, 15,000 feet — occasionally we'd get up to 16,000 feet with no oxygen. It gets pretty thin if you're up there too long. In the waist position, it would get very cold, particularly with the doors open.

"The waist position was right under the top turret. It was equipped a small piece of armor plate, but was only good if you got hit from directly underneath. I can't recall if we had just dropped our bombs or were still on the way in, but I was kind of curled up on the armor plate trying to get as much of me covered as I could. An Me 109 passed right over us and the top turret gunner was on him blasting away. As the 109 passed overhead, the turret spun around, and as he

THE BELL REAR DEFENSE ADAPTER, SEEN THROUGH A MAZE OF AMMUNITION FEEDS. THIS SYSTEM WAS USED IN BOTH THE B-25 AND B-26. THE TWO .50-CALIBER GUNS WERE FITTED TO A POWERED MOUNTING AND MOVED BY THE CONTROL HANDLES. THE GUNNER SAT BEHIND AN ARMORED SCREEN WITH A LIMITED VIEW THROUGH THE ARMORED GLASS. USAF MUSEUM, DAYTON.

did I felt something hit me in the back and I figured, 'Oh boy, I've had it.' I could feel this heat on my back and thought, 'This is going to hurt in a little while so I better get myself steeled up.' All of a sudden, the Me 109 spun around and came back, and when he did, the same thing happened. At the time, I had on this big flight jacket with a big fur collar. As the top turret opened up, I noticed the hot, spent shell casings falling from the turret and dropping down the back of my jacket. I thought for sure I had been hit by a flak burst.

"About the only time I felt I was really doing anything as a gunner was when we flew into a place we called 'Flak Alley' near Bezerti. It was a low-level mission, but we were right on the deck. I could man only one gun at a time, but I got some good shots at a bunch of bunkers. I felt I was accomplishing something, because I could see where the tracers were going. I'm sure somebody down below knew we were going over."

MARVIN YOST, B-26 FLIGHT ENGINEER / RADIOMAN / AIR GUNNER, 443RD SQUADRON, 320TH BOMB GROUP

THE VENTRAL BEAM POSITION WAS ADDED WITH THE INTRODUCTION OF THE B-26B MODEL.
TWO GUNS WERE PROVIDED, ONE ON EACH SIDE OPERATED BY A SINGLE GUNNER.
USAF MUSEUM, DAYTON

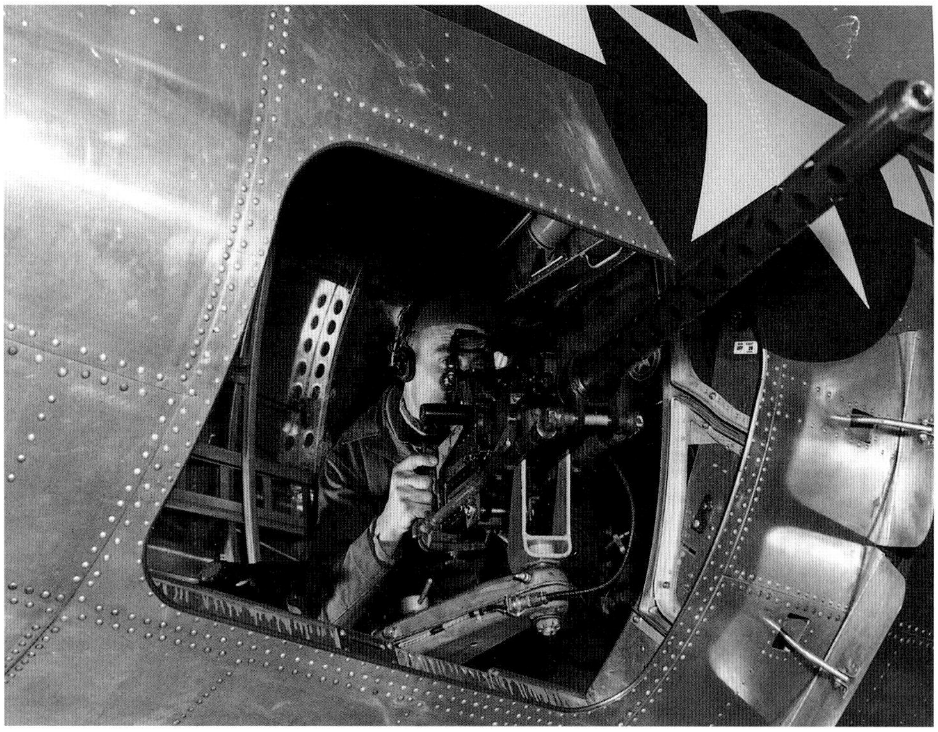

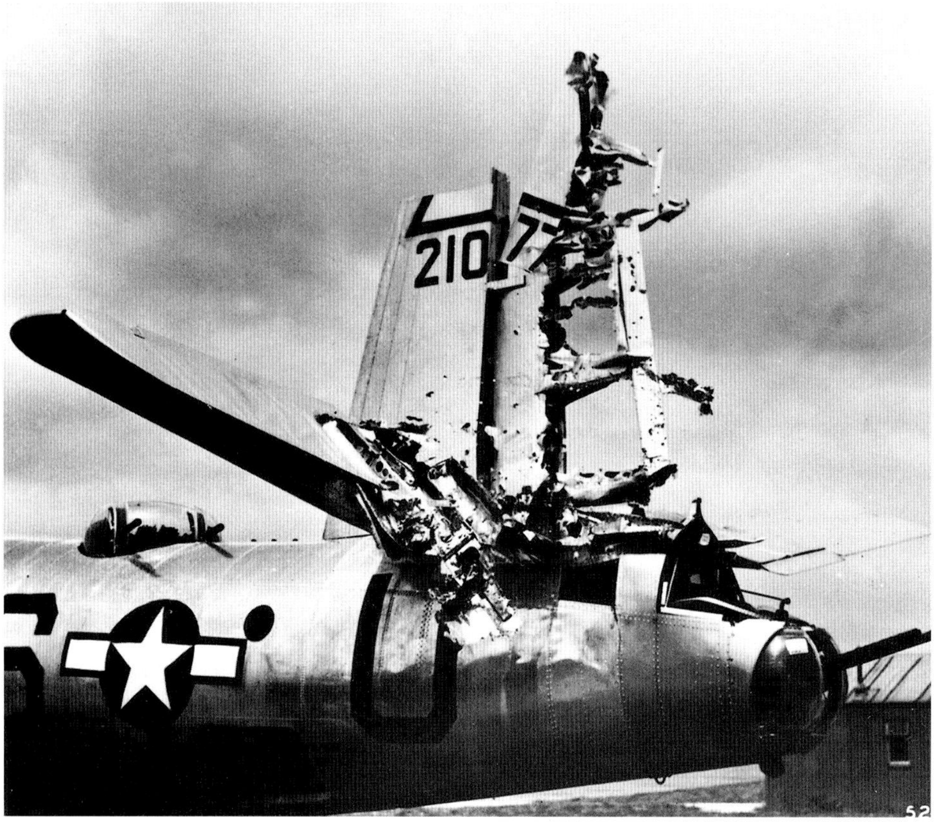
210

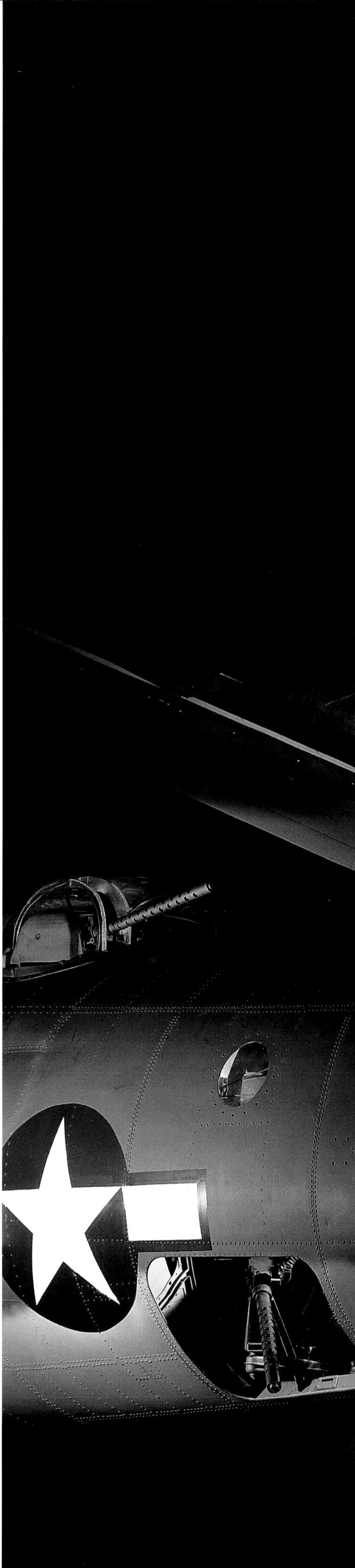

Top: A USAAF gunner takes aim. Note the wind deflector plates extended.

Above: Gunners were helpless against flak.
This heavily damaged B-26 was lucky to make it back.

Right: The B-26's defensive armament was concentrated in a very small space in the rear of the aircraft.
It had Six .50-caliber machines guns in all: two in the Martin top turret, two in the tail turret, and one each in the ventral positions.
USAF Museum, Dayton.

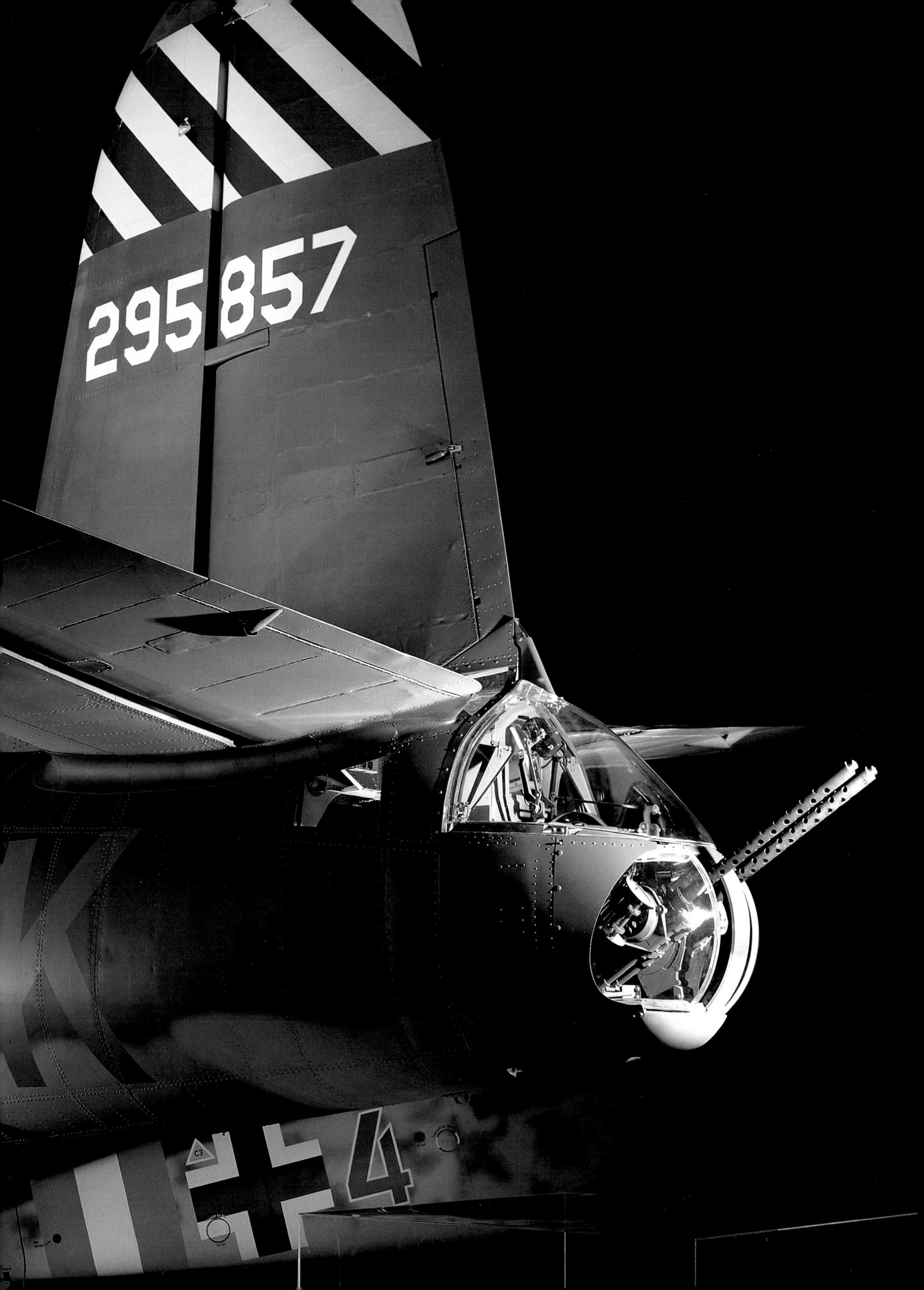
295857
4

Douglas A-20
Havoc

Compared to the B-25 Mitchell and B-26 Marauder, the A-20 was bereft of defensive armament, but it was a highly regarded and successful light bomber.

The Douglas A-20 was one of the last American bombers to receive a power turret. Early British versions were equipped with one and, later, two .303-caliber hand-held machine guns in the dorsal gun position and one .303-caliber gun in the ventral position. There was no flexible gun mounted in the nose. Forward armament consisted of two .303-caliber fixed guns mounted on each side of the fuselage just below the bombardier's position. Later American versions were up-gunned with .50-caliber machine guns. It was not until the introduction of the G model in 1943 that the Martin power turret was added. This turret was armed with two .50-caliber Colt-Browning machine guns and substantially increased the defensive firepower. This new turret provided exceptional all-round visibility for upper and rearward defense. There was a significant blind spot directly under the aircraft, but this was covered, at least in theory, by the lower ventral position. The addition of the Martin turret required extra room, so the fuselage was widened to accommodate the whole turret mechanism.

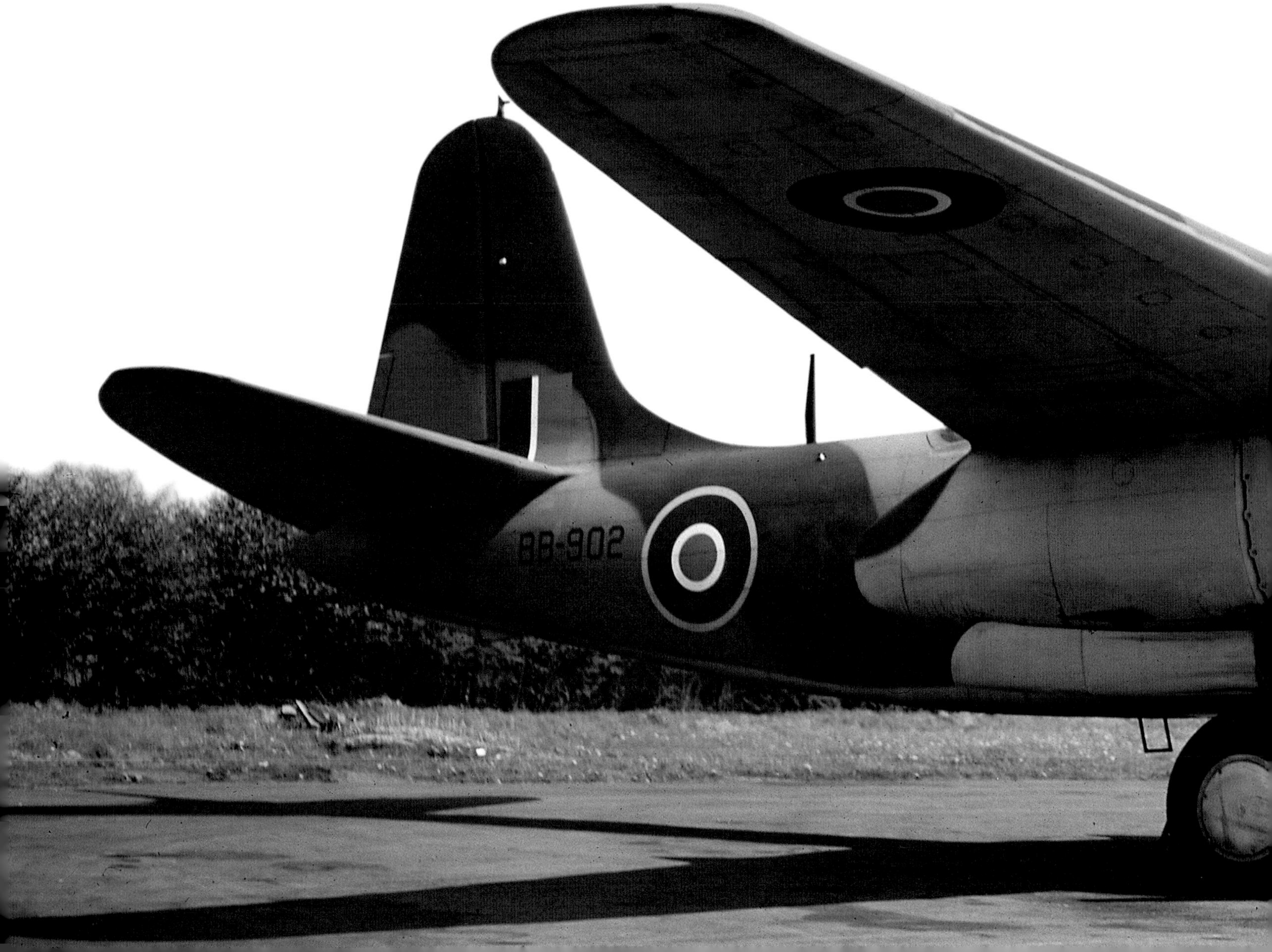
BB-902

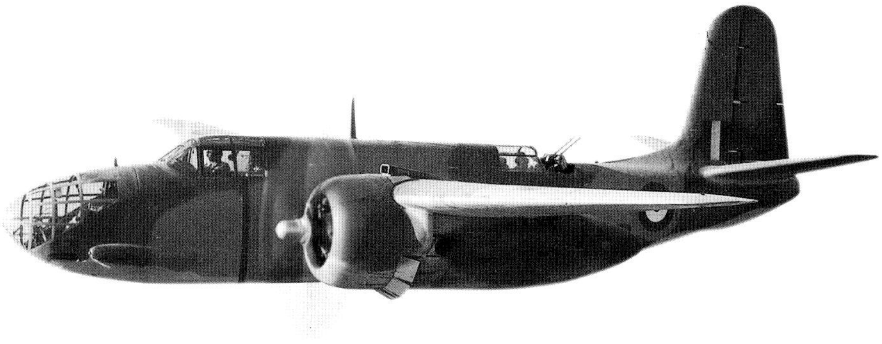

Eventually, 2,850 A-20Gs were produced, the most of any in the Havoc series.

Havocs first saw action with the French Air Force during the Battle of France in May 1940. Only sixty-four were in service when the Germans launched their attack through the Ardennes forest. These aircraft were committed piecemeal and had little effect on the German advance. After the fall of France, the remaining aircraft intended for the French were pressed into RAF service. There they operated in the night-fighter, intruder role. Later versions served with the Desert Air Force in the Africa, Sicily and Italy. In the Pacific, the A-20 was used in the low-level bombing role and was extremely effective. Solid-nosed A-20s with six .50-caliber machine guns proved highly effective against ships and airfields. In early 1944, three A-20 bomb groups were assigned to reinforce the 9th Air Force in preparation for the invasion of Europe. From their bases in England, A-20 crews bombed rail yards, airfields and troop concentrations in France, Belgium and Holland. After the invasion, they went ashore and provided close tactical support for the advancing American armies. Although lightly armed as compared to the B-25 and B-26, the A-20, like other medium bombers, operated for the most part with fighter escort. Early attempts to use the A-20 in the low-level bombing role without fighter escort proved too costly. In the end, a total of 7,385 A-20 Havocs were built, 3,125 of them going to Russia.

Above: The Boston III of the South African Air Force over the Libyan Desert.
Defensive armament consisted of two .303 machine guns in the dorsal position and one ventral gun.

Below: In 1940 the RAF ordered 300 Douglas DB-7s. Defensive armament consisted of a single .303 Browning machine gun in the dorsal position and one .303 machine gun in the ventral position. This DB-7 was used a trainer.
The DB-7 became the A-20 and was called Havoc by the Americans and Boston by the British.

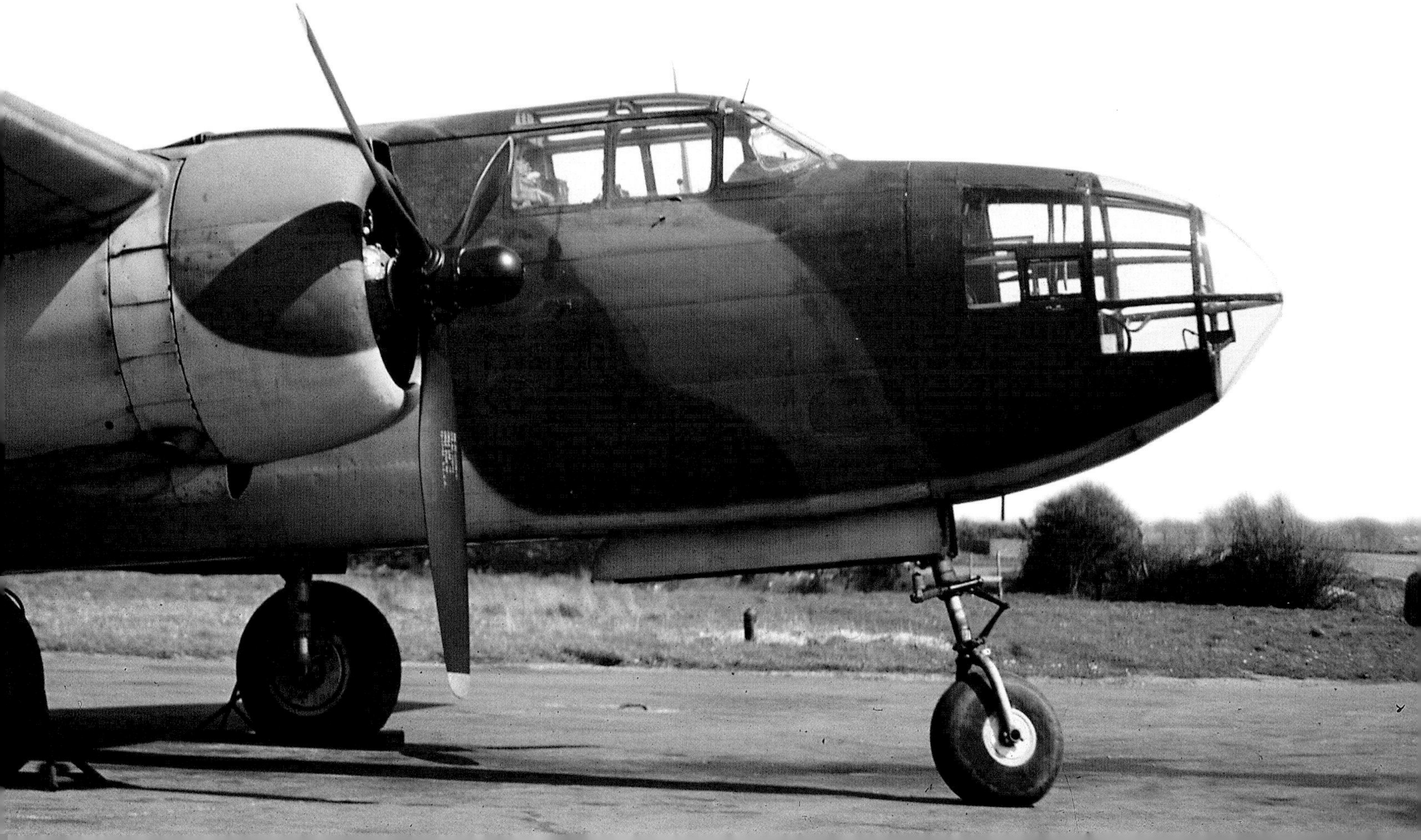

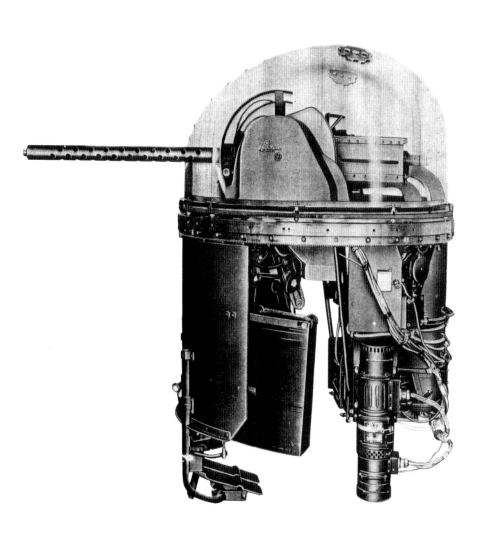

REPORT FROM AIR FIGHTING DEVELOPMENT
UNIT, DUXFORD 1941
WIRELESS OPERATOR /AIR GUNNER'S COCKPIT

The cockpit is roomy for its single occupant, giving a good view aft and forward
in the upper hemisphere, but restricted in the lower hemisphere by the main-
planes forward and the tailplane aft. The use of the upper gun is satisfactory
except that the perspex head does not form a windshield when opened. The
lower gun, however, can only be mounted after the hatch has been opened.
Leaving the gun mounted with the hatch permanently open makes the cockpit
draughty and cold. In action this is liable to be a great disadvantage.

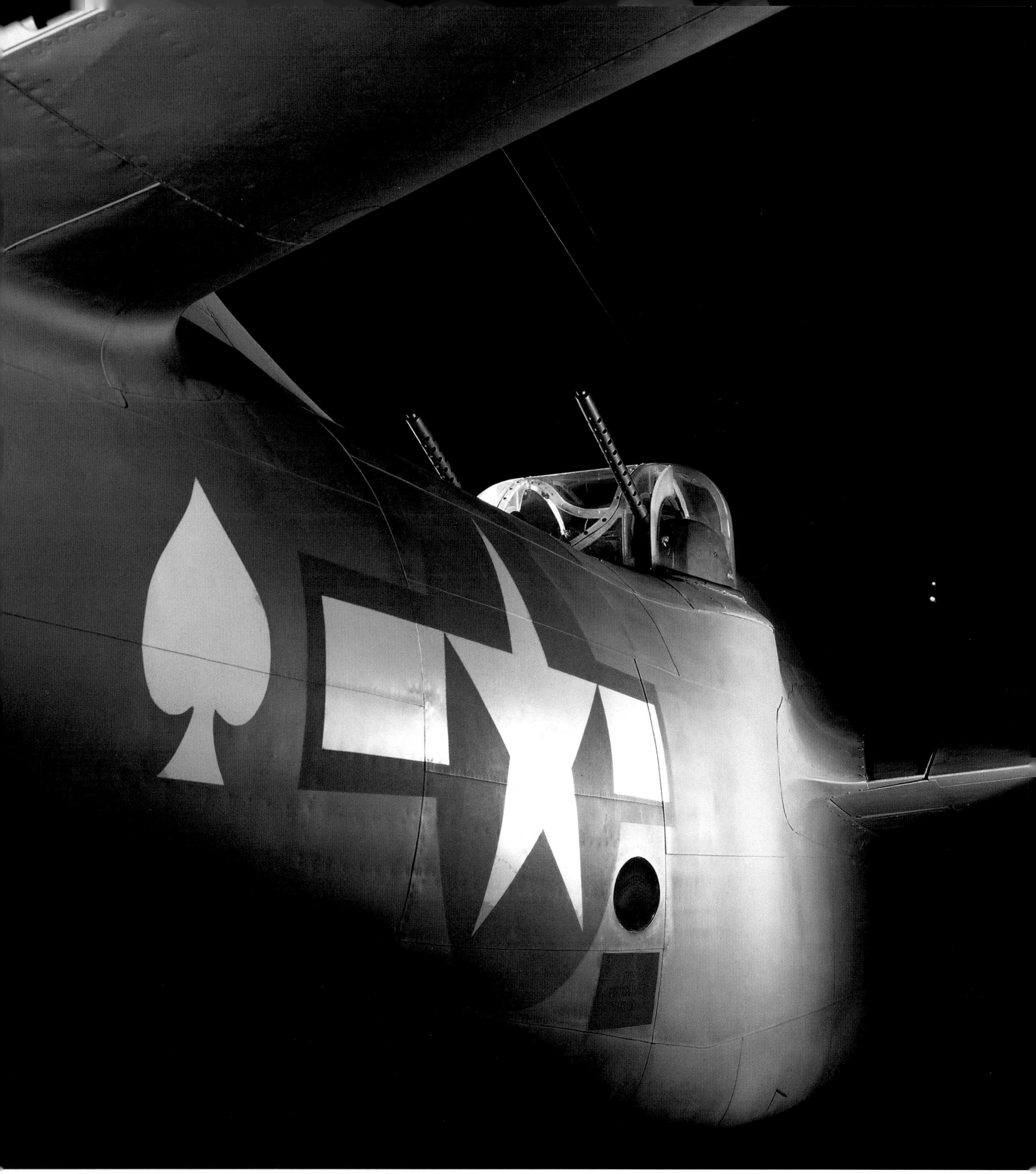

Above: The introduction of the Martin turret required a slight modification to the A-20. The A-20G's fuselage was slightly bulged in order to accommodate the Martin turret. The new turret provided excellent all-round view and substantially increased the A-20's rearward defense.
USAF Museum, Dayton.

Left: Detailed view of the Martin 250 upper turret. The ½ in (12.7 mm) armored apron gave the top turret gunner more protection from enemy fire than any other crew member.

Consolidated PBY
CATALINA

The Catalina rarely encountered enemy fighters. Its gunners fought boredom more often than they exchanged shots with the enemy.

Long-range maritime patrol was the least glamorous of combat duties. It involved flying over vast featureless oceans searching for ships and enemy submarines. Most aircrews never saw a thing, never fired their guns or dropped any bombs. Theirs was a war spent in lonely isolation. It was also a war that was vital for victory. It was the kind of warfare that required an aircraft of great range and rugged reliability, an aircraft such as the famous Catalina PBY.

The Consolidated Catalina first flew in 1935, and a contract for fifty aircraft soon followed. At the time of Pearl Harbor there were three squadrons of PBYs in US Navy service. The first shots fired by US Forces in World War II were delivered thanks to the Catalina. On December 7, before the Japanese attack on Pearl Harbor, a PBY crew spotted a periscope near the harbor entrance. The crew marked the sighting with smoke and guided the destroyer USS *Ward* to the scene. The destroyer quickly sank the sub and was credited with firing the first US shot of World War II.

BELOW: CATALINAS AND VENTURA BOMBERS OF FLEET AIR WING 4 AT THE READY IN THE ALEUTIANS, DECEMBER 16, 1943. BY AUGUST, ALL ENEMY ACTIVITY IN THE ALEUTIANS HAD COME TO AN END AND THOSE CATALINA CREWS LEFT WERE SUBJECT TO ENDLESS OCEAN PATROLS AND MIND-NUMBING BOREDOM.

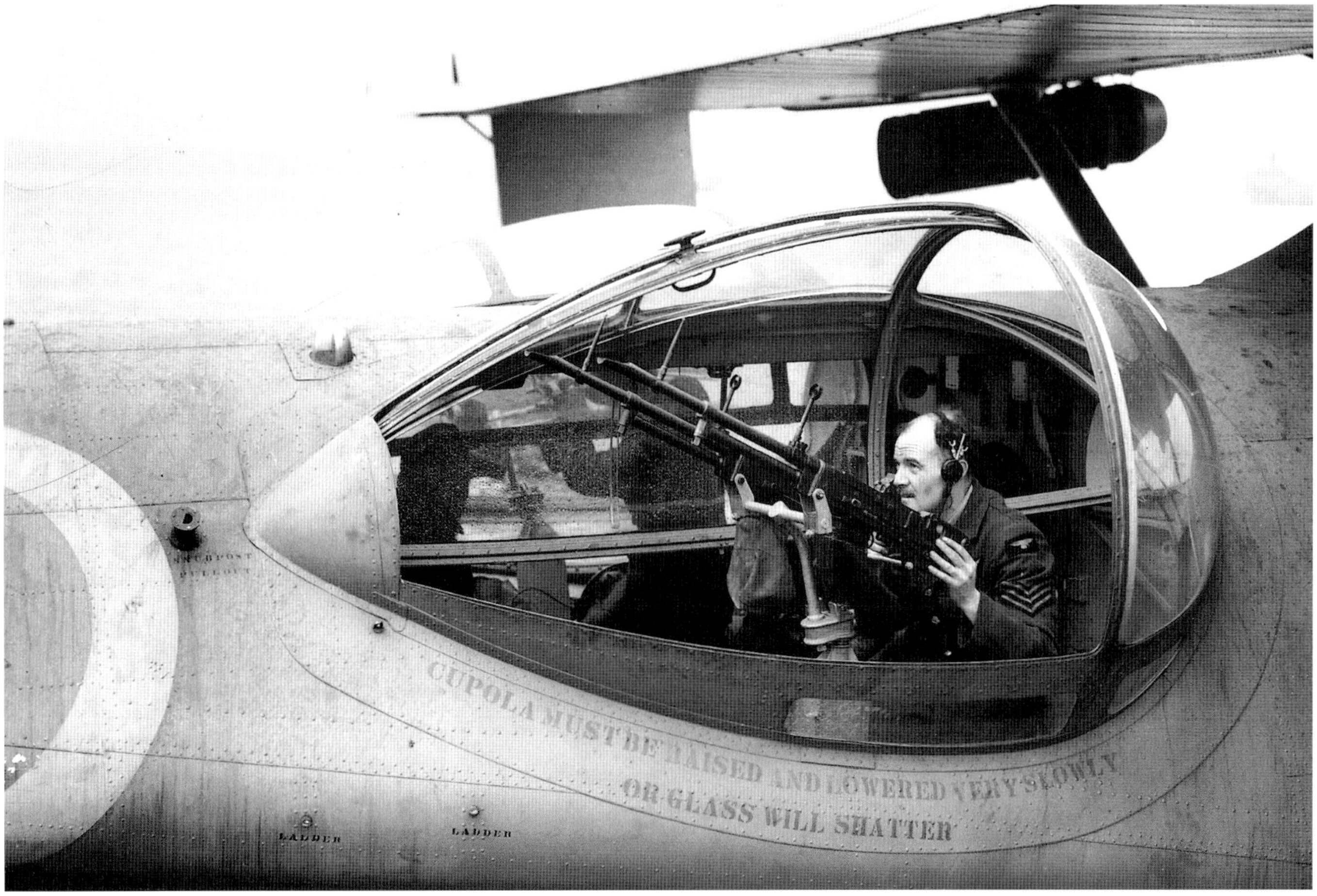
CUPOLA MUST BE RAISED AND LOWERED VERY SLOWLY
OR GLASS WILL SHATTER
LADDER
LADDER

The Catalina was a slow, steady aircraft. In fact, it was one of the slowest combat planes of World War II. Its armament consisted of one .50-caliber machine gun in each waist blister, one or two .30-caliber in the bow turret, and one .30-caliber machine gun in the rear ventral position. The Catalina rarely encountered fighters; it used its defensive armament more for offensive actions. By 1943, German U-boats were bristling with anti-aircraft guns. When German U-boats were caught on the surface, it was the job of the gunners to silence the anti-aircraft fire put up by the submariners. This was an extremely hazardous task, especially given the Catalina's slow speed. In all, over 3,000 Catalinas were built in the United States, Canada and the Soviet Union.

ABOVE LEFT: JESSE RHODES WALLER, AOM THIRD CLASS, TRIES OUT A .30-CALIBER MACHINE GUN HE HAS JUST INSTALLED ON A NAVY PBY.

ABOVE: THE CATALINA'S SIDE BLISTER GUN ARMAMENT VARIED THROUGHOUT THE WAR. HERE, SERGEANT A. SMITH MANS HIS TWO VICKERS K GAS-OPERATED GUNS.

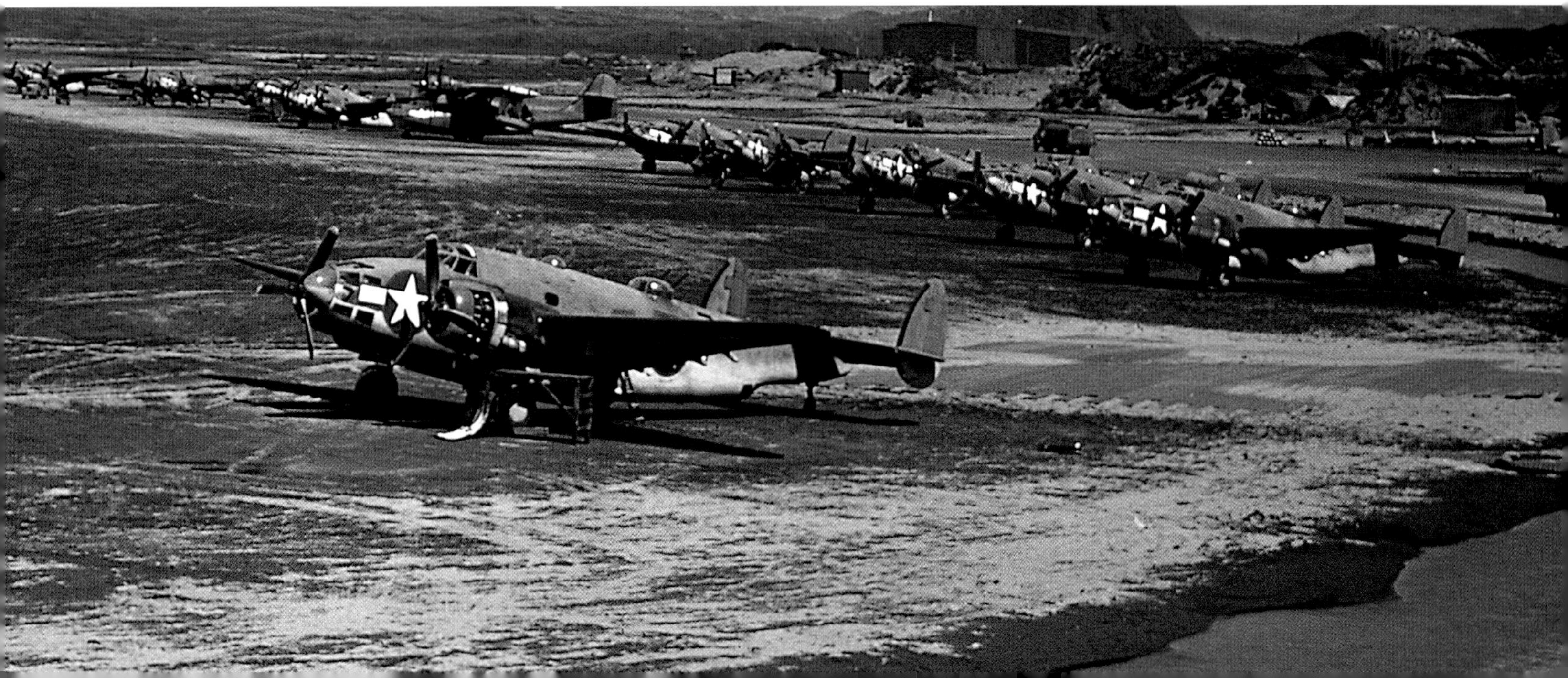

"The beautiful Cat — made to fly forever. Twenty-two hours at a time. Rest an engine, feather the prop, fly on one and then rest the other. The Squadron CO was my pilot. The Catalina had two big beautiful blisters, and for me, flying out of Mobasa (East Africa) into the India Ocean, stripped to my shorts, sunbathing on the gun deck, was a blast. There were no enemy aircraft in the Indian Ocean or the Gulf of Aden, just one or two subs to look for.

"Once, when I was with 230 Squadron on Sunderlands flying out of Gibraltar for Malta, a Cat came in on its last gasp of gas. Both its air gunners were missing. Their chutes, clothes, Mae Wests were all still in the aircraft. Someone had gone aft and had seen them both looking for enemy subs and aircraft at their respective blister positions. Later, when the same fellow returned, there was no sign of them. They searched for as long as they could. No one had a clue as to when, where, why or how they became missing.

"The Cat did a great job in the Indian Ocean, but we gunners were just an extra pair of eyes to look for subs and convoys. I never fired my guns in anger aboard the Cat, but what I can say is I believe the Cat was a superb platform."

FREDERICK "GUS" PLATTS, AIR GUNNER,
230, 282 AND 280 SQUADRONS

BELOW: THE PBY-6 WAS CONSIDERED THE BEST OF ALL VERSIONS OF THE CATALINA. THIS MODEL INCREASED ARMAMENT TO TWO .50-CALIBER MACHINE GUNS IN THE SIDE BLISTERS, ONE .30-CALIBER GUN IN THE VENTRAL POSITION AND TWO .30-CALIBER GUNS IN THE NEW, EYEBALL FRONT TURRET.
NATIONAL MUSEUM OF NAVAL AVIATION, PENSACOLA.

RIGHT: CLOSE-UP OF THE CATALINA SIDE BLISTER. THE .50-CALIBER MACHINE GUN PROVED TO BE THE MOST SUCCESSFUL AIRCRAFT GUN EVER PRODUCED. THE STANDARD M2 WEIGHED 64 POUNDS AND FIRED A BULLET THAT WEIGHED 1.7 OUNCES TO A MAXIMUM RANGE OF 21,600 FEET. RATE OF FIRE WAS BETWEEN 750 AND 850 ROUNDS PER MINUTE.
USAF MUSEUM, DAYTON.

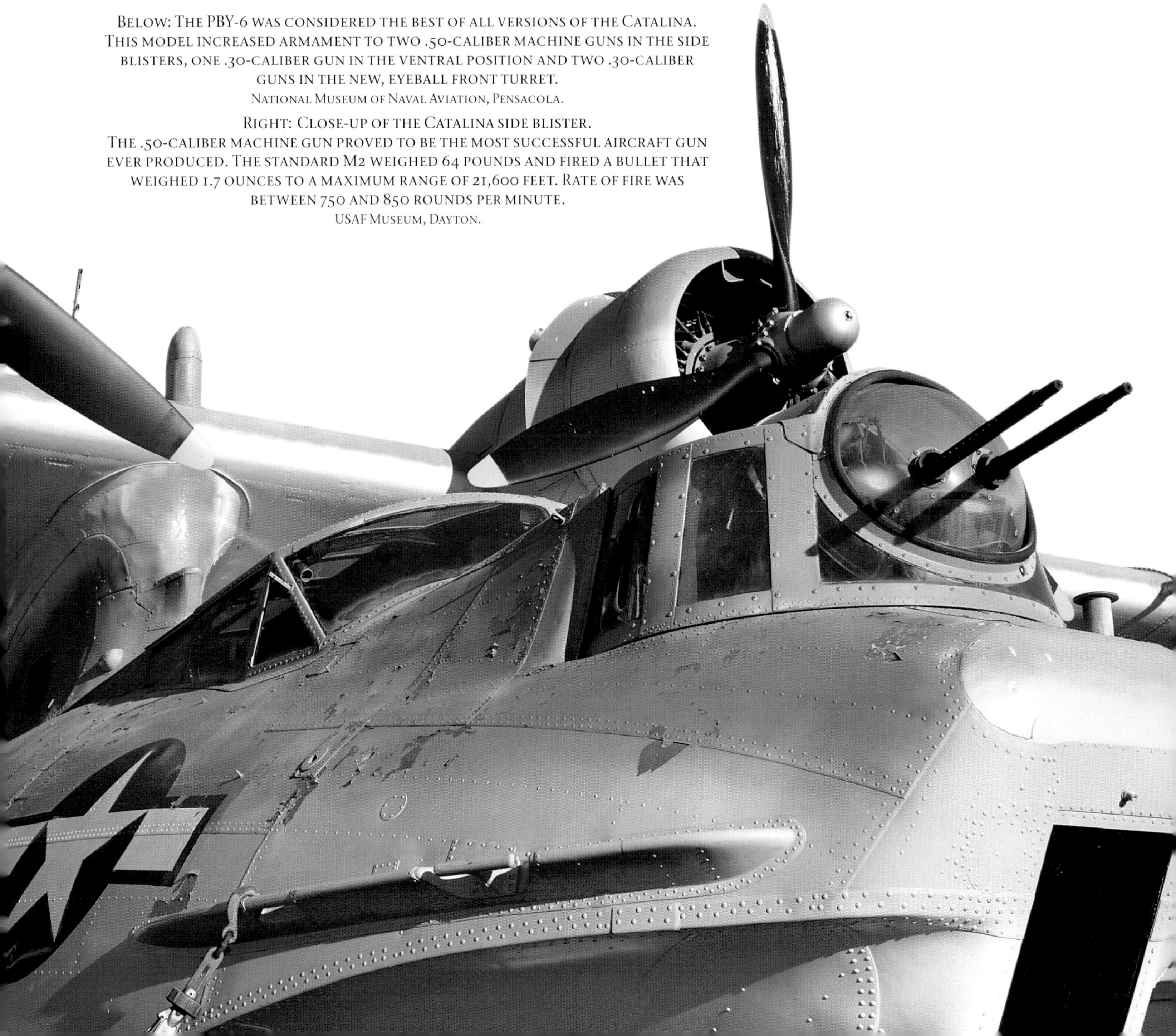

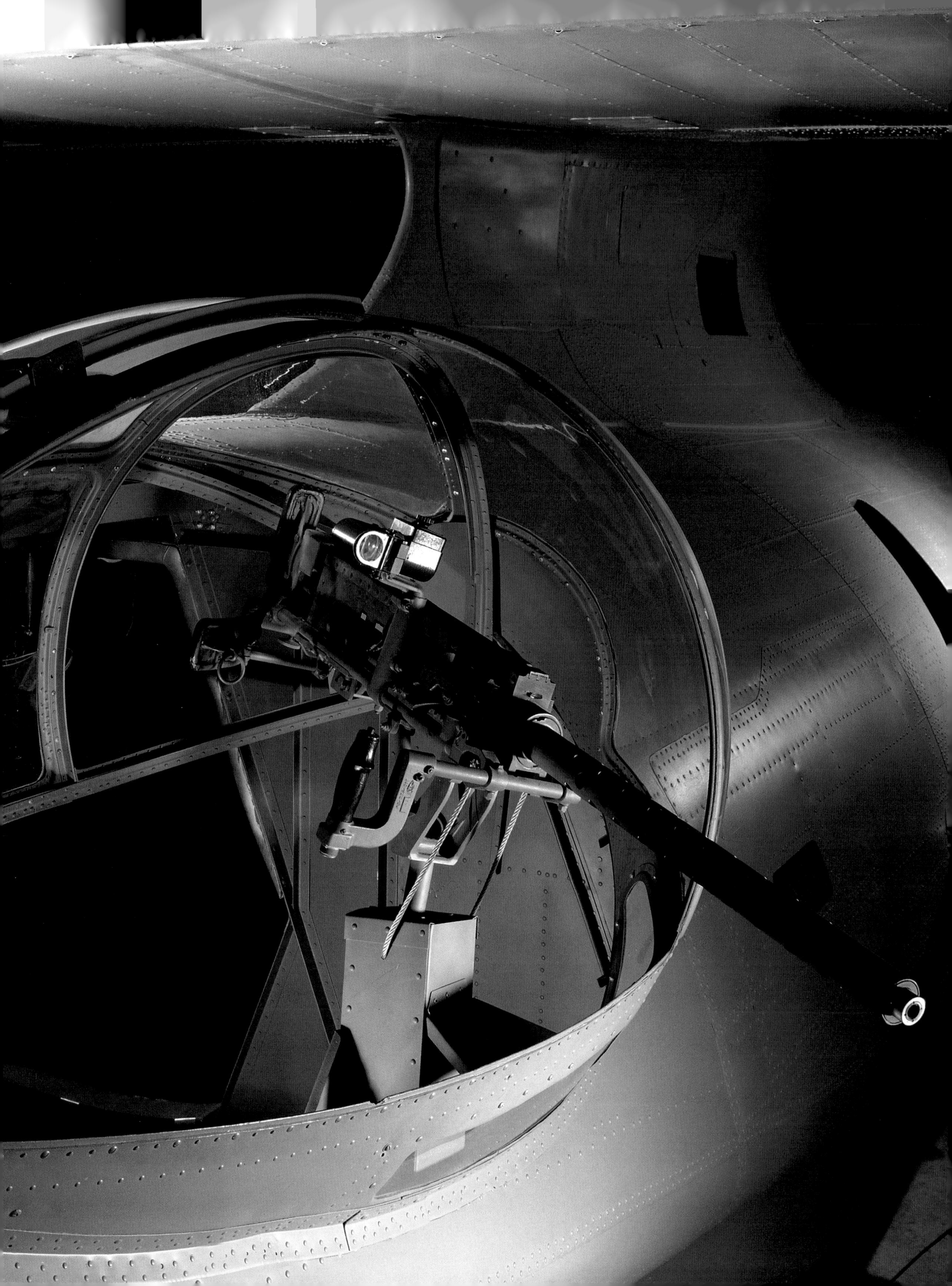

BRISTOL
BLENHEIM

ALTHOUGH THE BLENHEIM NOTCHED MANY WARTIME FIRSTS FOR THE RAF, IT WAS ALWAYS
LIGHTLY ARMED AND SUFFERED HEAVY LOSSES WHEN NOT PROVIDED
WITH LONG-RANGE FIGHTER ESCORT.

When the Bristol Blenheim was first displayed at the RAF's Hendon display in the summer of 1937, it caused much excitement. Its high speed and modern appearance convinced many that the RAF was armed with the world's most advanced and formidable bomber aircraft. The Blenheim never did live up to the hype. Blenheim-equipped daylight bombing squadrons suffered appalling losses in the early stages of World War II. Armed only with a single .303-caliber machine gun in the wing and a Bristol Type B.I powered turret with a single Lewis gun, the Blenheim was no match for the cannon-armed fighters of the Luftwaffe.

The Battle of France exposed the Blenheim's many weaknesses. Gunners complained of the lack of armor protection and the pathetic firepower performance of the single Lewis gun. As a stopgap measure, armor was fitted in the rear of the fuselage behind the turret and the firepower was increased to two belt-fed Browning .303-caliber machine guns. Even with the increased punch of two machine guns, the Bristol turret was outdated and of little use. Later versions of the Blenheim were equipped the more advanced Bristol Type X upper turret, along with the Frazer Nash FN 54 under defense turret. The Type X was a well-designed and armored turret, but it was still no match against well-flown heavily armed fighters. The Frazer Nash FN 54

turret was mounted under the nose and operated by the bomb aimer using a periscope sight. It was designed to counter fighter attacks from below and behind, but was never considered effective. The 20-degree field of vision made it extremely difficult to locate an attacking fighter, especially when the pilot was using evasive action. In Europe, the Blenheim was finally replaced in 1942 by Douglas Bostons and de Havilland Mosquitos. In the Far East, it would serve until 1943 without distinction.

"The Bolingbroke [Canadian-manufactured Blenheim] was equipped with a Bristol turret, and I found that its movement was somewhat jerky. I didn't think it was a very good turret compared to the others we trained on during turret manipulation."

FLYING OFFICER DOUG SAMPLE,
415 SQUADRON, 6 GROUP, RCAF

"We flew back to England from France on December 1, 1939, to re-equip on Blenheims. Compared to the Battle, the Blenheim was a Rolls Royce. The Blenheim turret was only armed with a single Vickers K gun. It was pretty cramped and I didn't like turrets myself. I wasn't used to confined spaces."

SERGEANT JOHN HOLMES,
AIR OBSERVER / GUNNER

ABOVE: NOT AS FEARSOME AS IT LOOKS. AN RAF GUNNER POINTS HIS TWIN .303 BROWNINGS AT THE CAMERA.
RIGHT: THE BRISTOL BLENHEIM MK IV F (FIGHTER VERSION) IN FORMATION WITH GUNNERS AT THE READY. THE FIGHTER VERSION
WAS USED FOR CONVOY PROTECTION AND LONG-RANGE PATROL DURING THE INVASION OF NORWAY.

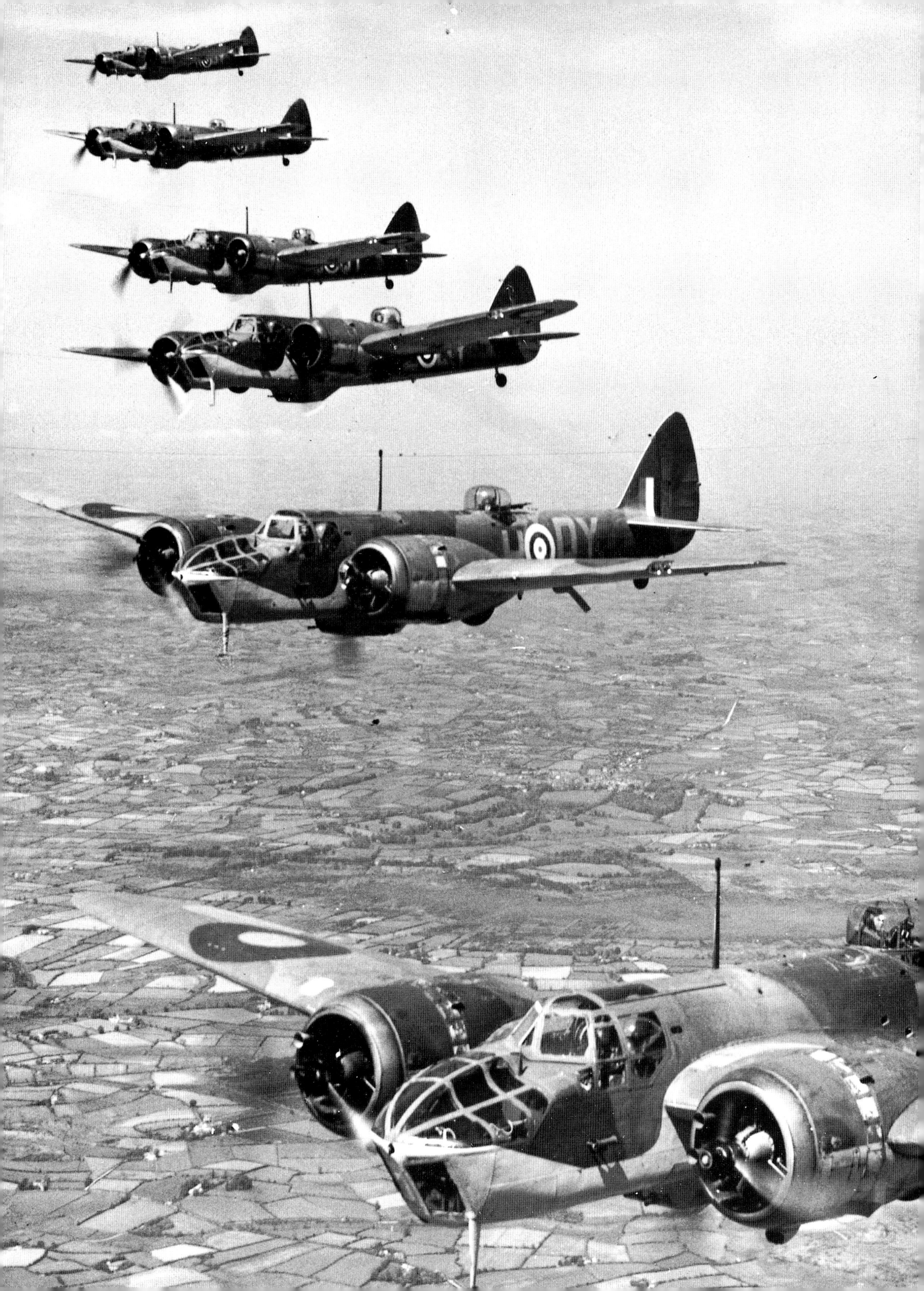

"During training we used twin-gun Blenheim-type turret (Bristol, I presume) fitted in Ansons, which were rather uncomfortable, but nevertheless quite maneuverable, and had the advantage of a retractable seat that rose and fell with the elevation of the guns. That meant a good line of sight was maintained between the eye, the gunsight and the target."

PETER CALDWELL, AIR GUNNER, RAF

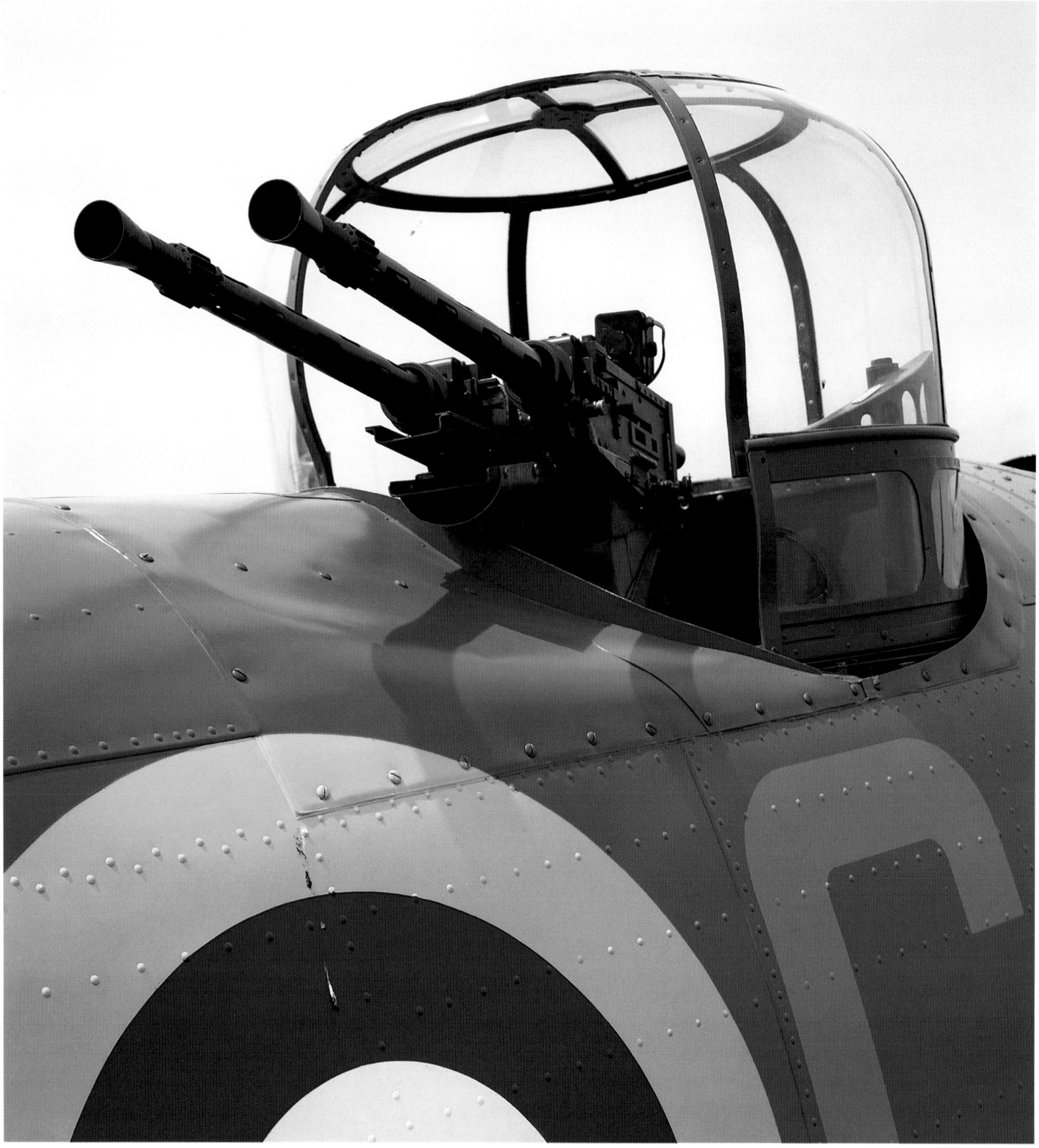

ABOVE: EARLY VERSIONS OF THE BLENHEIM WERE ARMED WITH JUST A SINGLE LEWIS GUN. THIS WAS LATER UP-GUNNED TO TWO .303 BELT-FED MACHINE GUNS. THE BROWNING MK II AS SEEN HERE WAS THE GUN FITTED TO MOST RAF TURRETS.
AIRCRAFT RESTORATION COMPANY.

RIGHT: THE INTERIOR OF THE BRISTOL TYPE B.I TURRET. AFTER ENTERING THE AIRCRAFT THROUGH A HATCH FORWARD OF THE TURRET, THE GUNNER FIRST STOWED HIS PARACHUTE ON A RACK BESIDE THE TURRET AND THEN STOOPED DOWN AND UP INTO THE TURRET.
AIRCRAFT RESTORATION COMPANY.

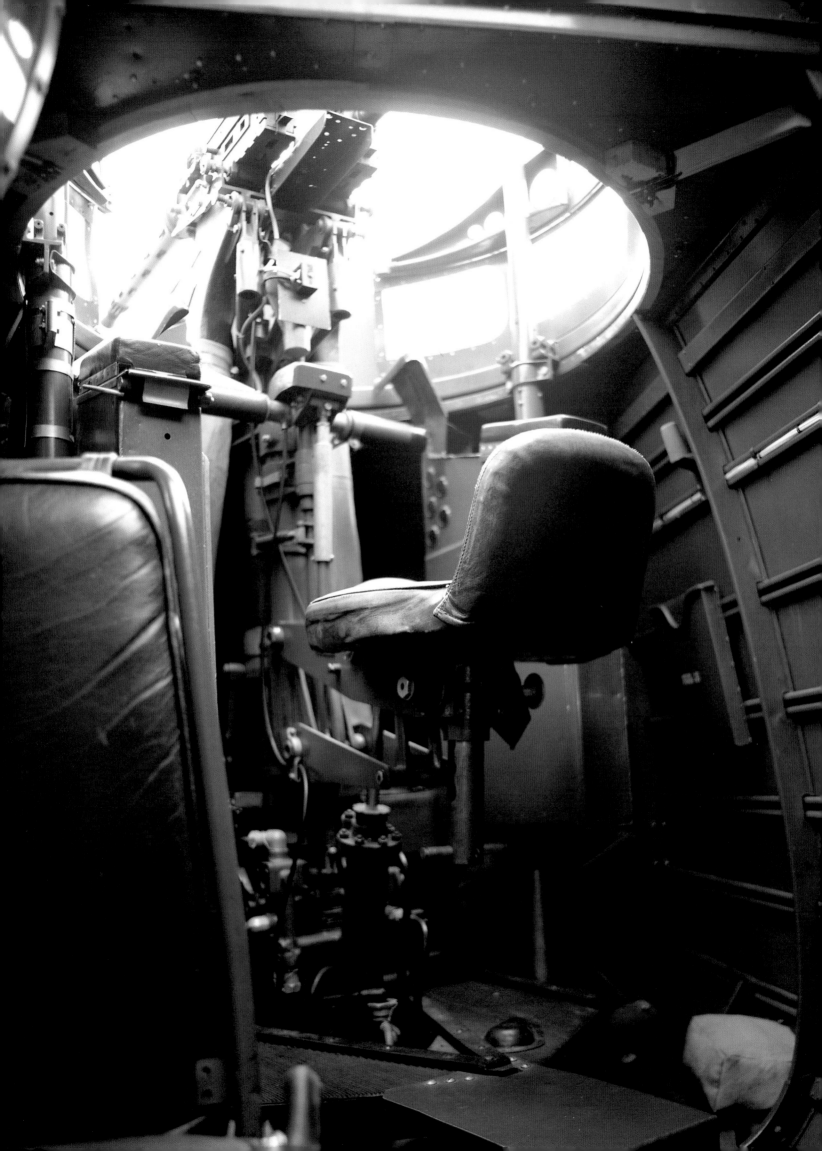

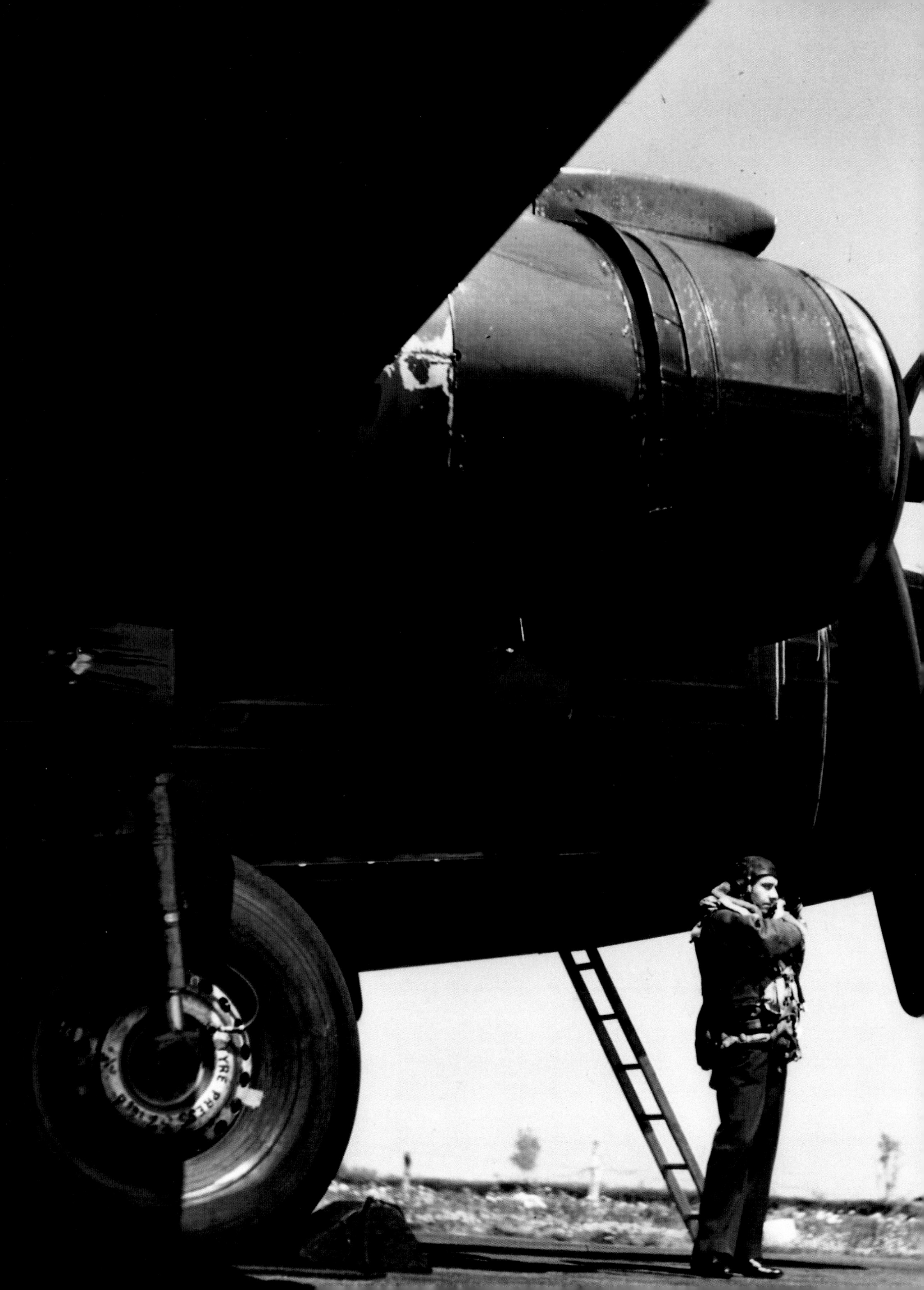

VICKERS
WELLINGTON

AT THE BEGINNING OF WORLD WAR II, THE WELLINGTON WAS ARMED WITH THE WORLD'S MOST ADVANCED POWERED TURRETS. EVEN WITH THIS ADVANTAGE, THE WELLINGTON SOON PROVED VULNERABLE TO DAYLIGHT ATTACKS AND WAS FORCED INTO THE NIGHT SKY FOR PROTECTION.

When war began, Bomber Command had thirty-three operational squadrons. Sixteen were classified as light and were equipped with the Blenheim and Battle; the other seventeen were listed as heavy squadrons and were equipped with the Wellington, Hampden and Whitley. These heavy squadrons had just over 200 operational aircraft, and by 1939, both the Hampden and Whitley were considered obsolete even by prewar British standards. Only the Wellington regarded as top of the line — and it was.

The Wellington was an outstanding design, and much of its success was due to the revolutionary geodetic system of construction. This system combined great strength with light weight. When the Wellington Mk I entered service, it was armed with two Vickers turrets. Each turret was equipped with a single Lewis gun. These turrets were not popular with the gunners because, even though they were very roomy, they offered poor sighting and gun control. After numerous reports of failures, the Vickers turrets were replaced with the Frazer Nash FN 5 turret. The FN 5 was mounted in both the nose and rear of the aircraft, but shortly after that, the FN 5 rear turret was replaced by the FN 20 four-gunned turret. The Wellington also held two .303-caliber Browning machine guns in the waist position.

The Wellington carried the lion's share of RAF Bomber Command's early night-bombing offensive and was still in front-line service when the war ended. Because of its unique construction, the Wellington was able to absorb a fantastic amount of battle damage and return its crews safely back to base. Although the Wellington proved the power-operated gun turret could be a credible defense, it also showed that large formations of bombers could not hope to survive against heavily defended targets without fighter escort.

A total of 11,461 Wellingtons were built. They served in Europe, Africa, Italy and the Far East in a remarkable diversity of roles. Other bombers flew higher and faster, but none enjoyed a greater reputation.

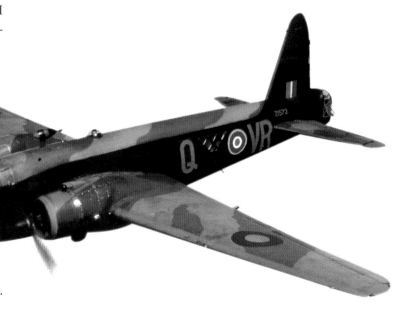

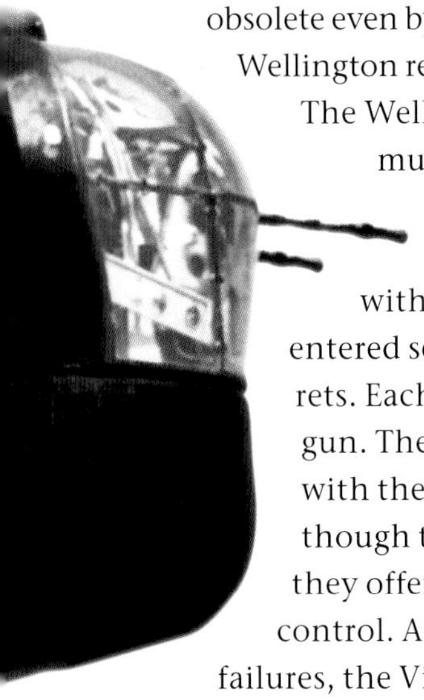

ABOVE: THE VICKERS WELLINGTON B. MK III WAS THE MOST HEAVILY ARMED BRITISH BOMBER IN THE EARLY STAGES OF THE SECOND WORLD WAR. IT CARRIED EIGHT .303 MACHINE GUNS; SIX IN TWO POWERED TURRETS AND TWO WAIST GUNS.

LEFT: AN RCAF CREWMAN PREPARES FOR ANOTHER FLIGHT. THE INTERIOR OF THE FN 5 FRONT TURRET IS NOT PAINTED BLACK, INDICATING THIS PHOTOGRAPH WAS TAKEN EARLY IN THE WAR.

"All told, I flew over 300 hours out of Malta on my first tour and over 300 flying out of England on shipping strikes and anti-submarine patrols.

"On my first tour, I flew as a rear gunner in Wellingtons out of Malta. We flew mostly at night against convoys. I never fired my gun in anger. You always had to keep in mind that the other guy is going to shoot back. We never saw a lot of German night fighters, but we were shot at many times by flak from enemy convoys. Once in a while you'd get a glimpse of a night fighter against a cloud bank. One night we were coming back from a mission and we were flying through a layer of clouds. It was also a moonlit night. I was in the tail turret. As we came out of some cloud, I happened to look

down and spotted a Ju 88 a couple of hundred feet below. It was so bright I could see the crosses on the wings. At this point I could have let him have it, but I told the pilot, and he said, 'Don't do anything.' We quickly climbed back into the clouds and escaped. If the Ju 88 pilot had looked up, I'm sure he would have seen us and I might be telling you a different story.

"I would have felt more comfortable with more armor plating in the rear turret. Everyone was issued with the standard steel helmet. I always sat on mine."

ERIC CAMERON,
WIRELESS AIR GUNNER,
69, 221, 407, 489 SQUADRONS

ABOVE: IN THE NIGHT AIR WAR OVER GERMANY, THE FRONT TURRET WAS NEXT TO USELESS. OVER 22,000 FN 5 TURRETS WERE PRODUCED AND USED IN THE WELLINGTON, STIRLING, MANCHESTER AND LANCASTER.
RAF MUSEUM, HENDON.

RIGHT: INTERIOR OF THE FN 5 TURRET. THE FN 5 WEIGHED 252 POUNDS AND THE TWO MK II .303 MACHINE GUNS WERE PROVIDED WITH 1,000 ROUNDS PER GUN.
RAF MUSEUM, HENDON.

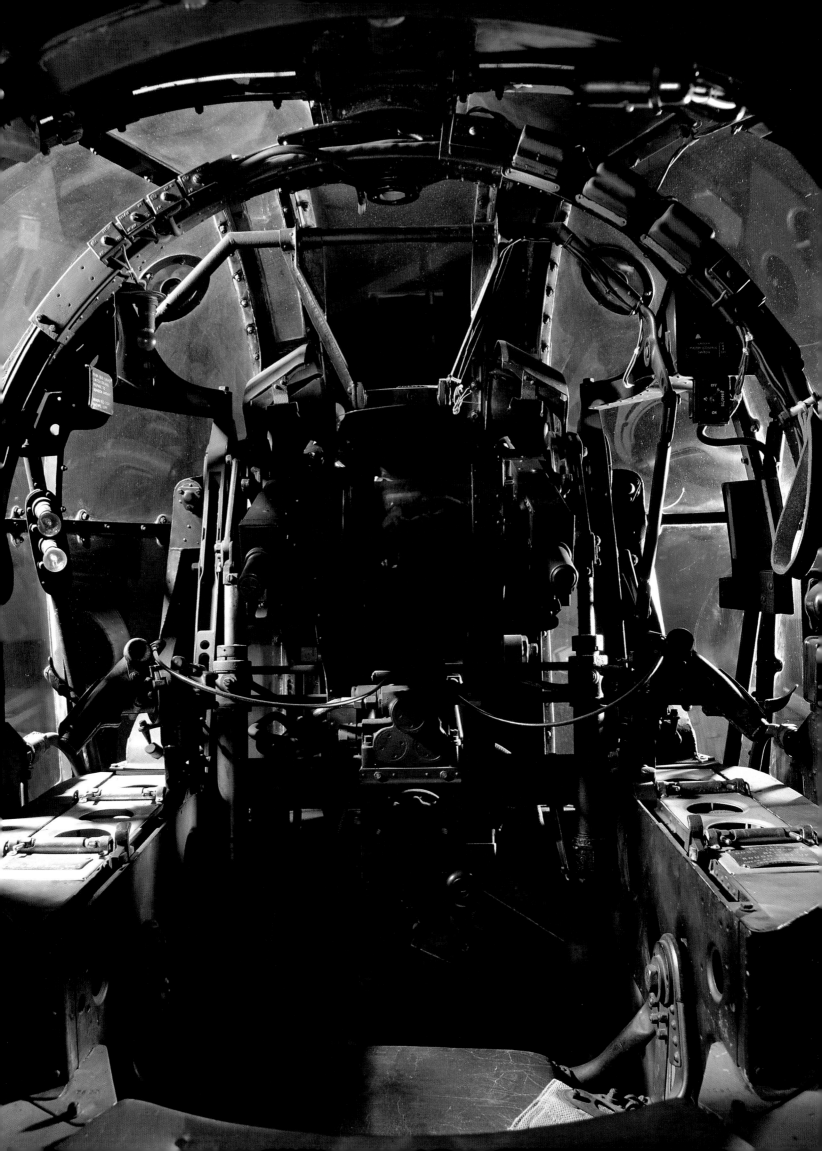

MESSERSCHMITT
BF 110

AS A LONG-RANGE ESCORT FIGHTER, THE BF 110 DID NOT LIVE UP TO ITS PROMISE, BUT AS AN
INTERCEPTOR PROWLING THE NIGHT SKIES OVER GERMANY, IT BECOME ONE OF THE MOST
SUCCESSFUL NIGHT FIGHTERS OF ALL TIME.

While the British and Americans were reluctant to consider the idea of a long-range fighter, the Germans fixed on the concept of a long-range strategic fighter. Lessons from the First World War had shown the need for a fighter with extended range and endurance, to provide bomber escort or to gain air superiority behind enemy lines. The first pre-production Bf 110 *Zerstörer* (destroyer) flew on April 19, 1938, and during the invasion of Poland, the Bf 110 would prove highly effective. The centralized armament of two 20-mm cannon and four 7.92-mm machine guns and the weight of fire from these weapons proved devastating. The Bf 110 was also equipped with a single 7.92-mm machine gun in the rear gunner's position.

It also became apparent that the Bf 110 could not match the performance of single-engine fighters. In fighter-versus-fighter combat, where snappy rates of roll, turning ability and acceleration win the day, the Bf 110 was outclassed.

Ironically, one of the most significant contributions made by the Bf 110 was to expose the use of the self-defending turret-armed bomber as inadequate and dangerously flawed. On December 18, 1939, in what has since been dubbed the Battle of the German Bight, twelve of twenty-two Wellington bombers sent to bomb Wilhelmshaven were shot down. Although the defending fighters were a mixed bag of Bf 109s and Bf 110s, it was the 110 that played a prominent role. Shortly after, British bombing policy changed, and the bomber offensive against Germany would be carried out only under the cover of darkness.

After the Battle of Britain, the shortcomings of the Bf 110 as a day fighter were clear, but as Bomber Command's night-bombing offensive increased, RAF gunners found themselves once again battling the Bf 110. As a night fighter, the Bf 110 adapted well, and it flew until the end of the war. Although it failed as a long-range escort fighter, the Bf 110 was one of the most effective and deadly night fighters of the war. The majority of Germany's highest scoring night-fighter aces gained most of the kills flying the Bf 110.

A LINE-UP OF BF 110S ABANDONED IN NORWAY
SHORTLY AFTER HOSTILITIES ENDED.

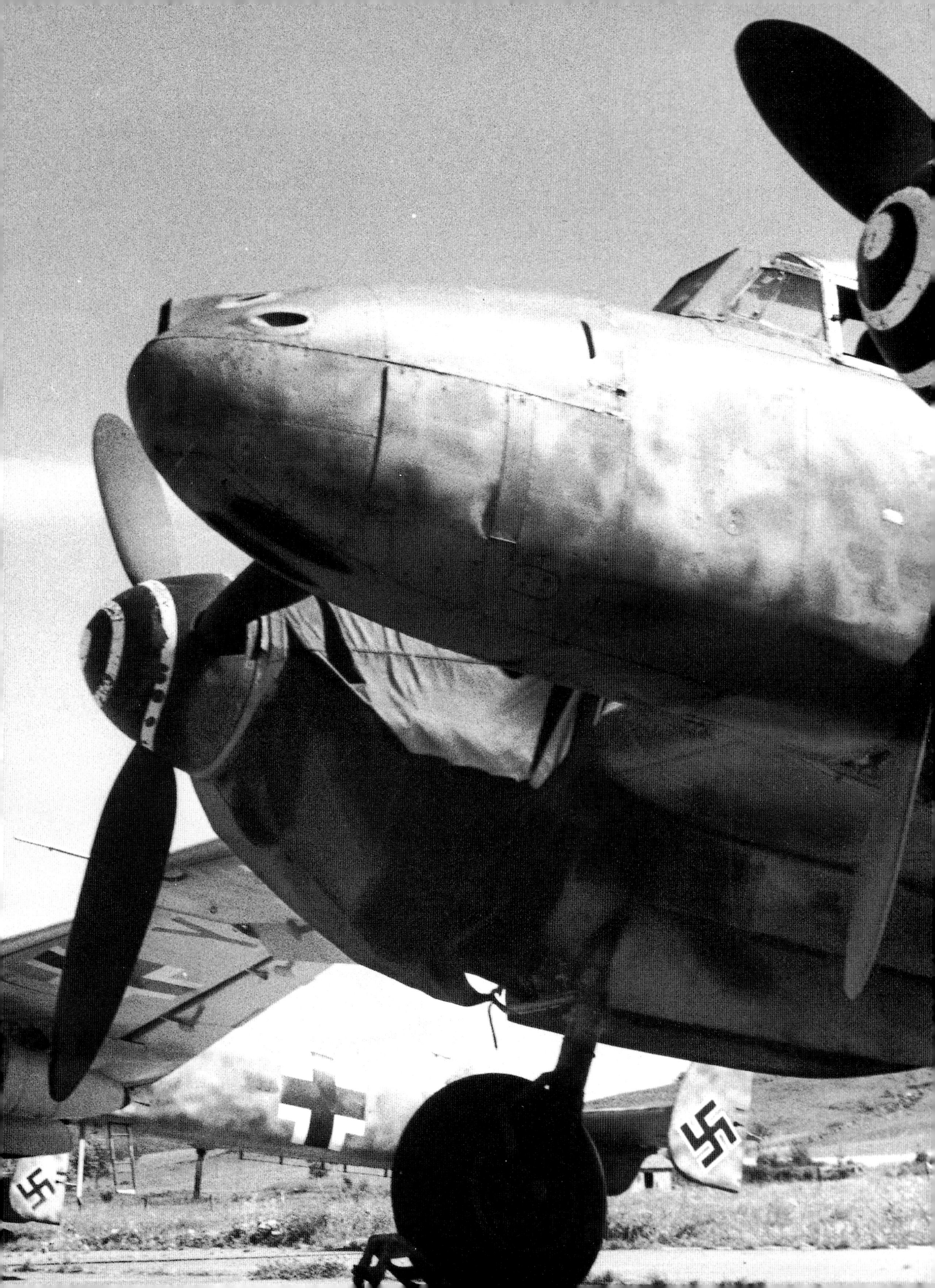

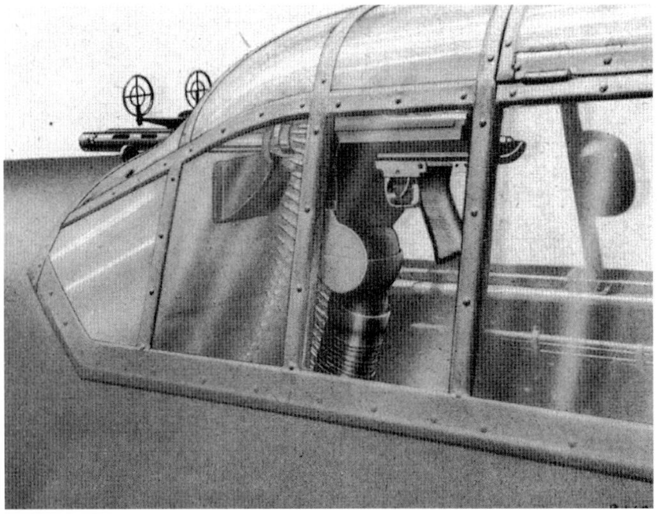

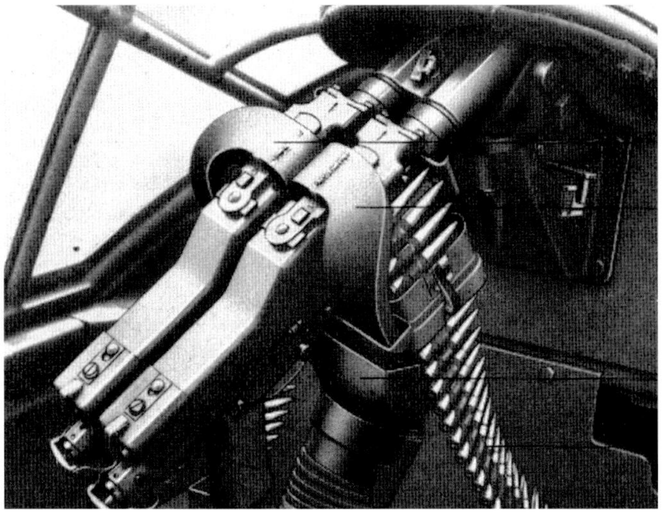

"As a radio operator / air gunner, I only flew at night. I began flying operations in 1943 with my pilot, Martin Becker. We flew the Messerschmitt Me 110 and later the Ju 88. Compared to my training as a radio operator, I received very little as a gunner. The training I did receive consisted mostly of shooting at ground targets; even that was difficult. At the time, we had only the single MG 15 7.62-mm machine gun mounted in the rear. The gunner's cockpit in the Me 110 was fully enclosed. After a while we received the newer version equipped with the twin MG 81Z Zwilling machine gun. This gun was cocked and fired electrically.

"The gunner's cockpit in the Me 110 was very narrow, especially when there were three people in the aircraft — the pilot, radio operator and gunner. When I flew as a radio/radar operator, the gunner and I sat with our backs against each other. I sat looking forward, behind the pilot. Operating the radar and radios was a big job, and in that situation, it was not practical for me to act as a rear gunner as well. The Me 110 was also equipped with two MG FF 20-mm cannons mounted in the cockpit floor underneath the radio/radar operator's compartment. This meant you could change the sixty-round drum magazines in flight. As you can imagine, this was a very difficult thing to do especially in the dark, not to mention in an aircraft turning and maneuvering in the night sky."

KARL JOHANSSEN, ME 110 / JU 88
RADIO OPERATOR / AIR GUNNER

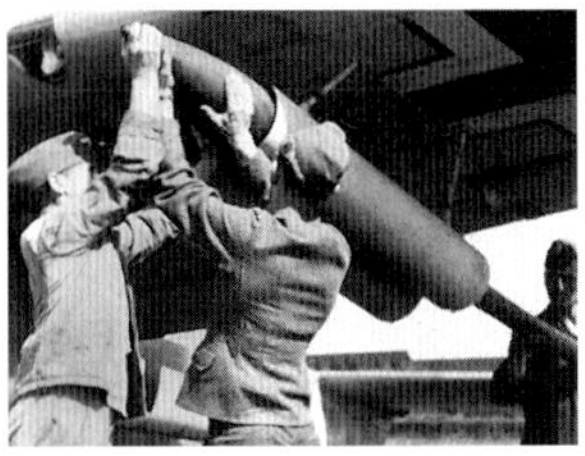

A

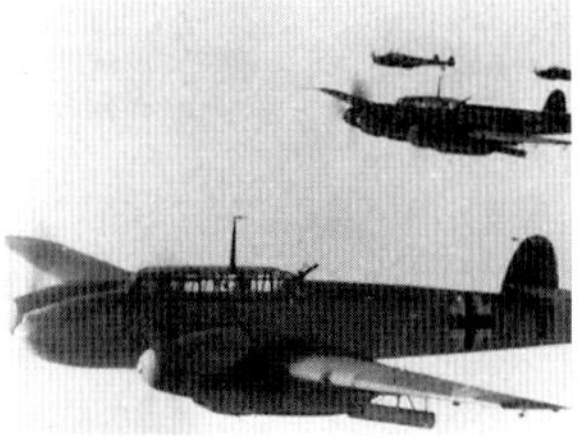

B

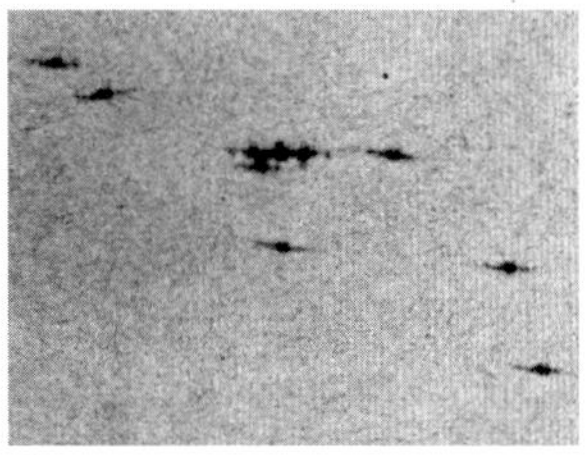

C

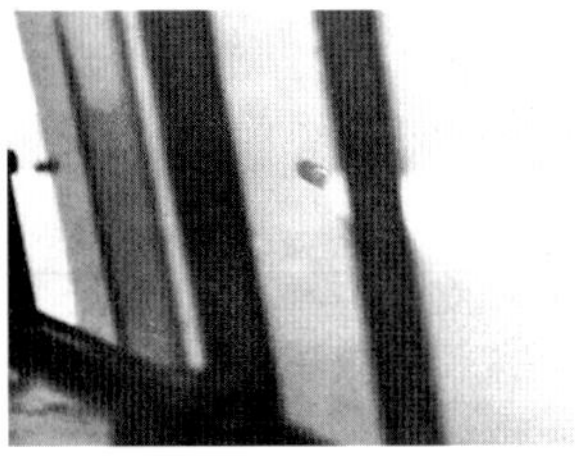

D

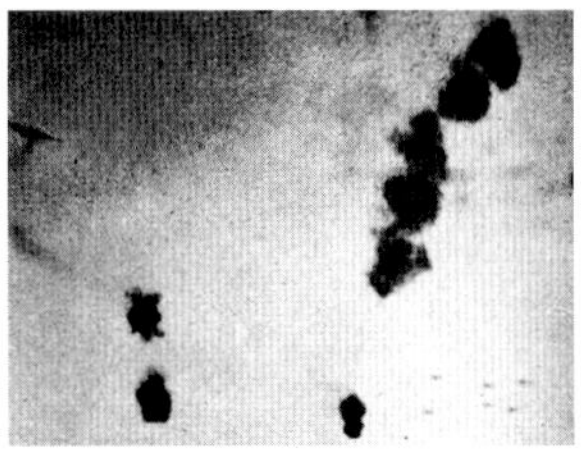

E

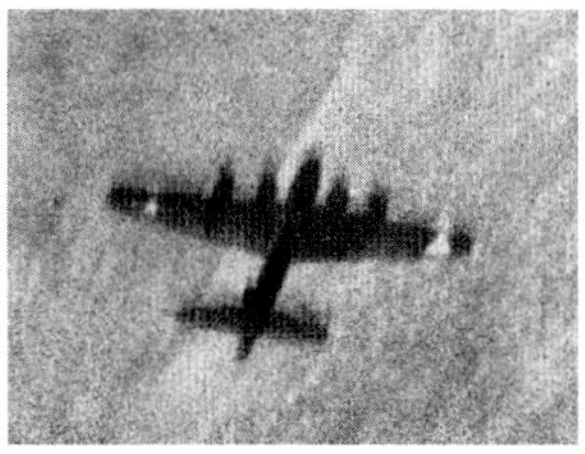

F

AN EARLY BF 110 AIR DEFENSE OPERATION UNFOLDS:

(A) A BF 110 IS LOADED WITH WGR 210-MM ROCKETS WITH THE APPROACH OF AMERICAN HEAVY BOMBERS.

(B) BF 110S FORM UP.

(C) CLOSING IN ON A LARGE FORMATION OF B-17S.

(D) UNDERWING ROCKETS ARE LAUNCHED.

(E) EXPLOSIONS ROCK THE B-17 FORMATION; ONE B-17 IS DAMAGED AND BEGINS TO TRAIL SMOKE.

(F) THE LAST MOMENTS OF A B-17.

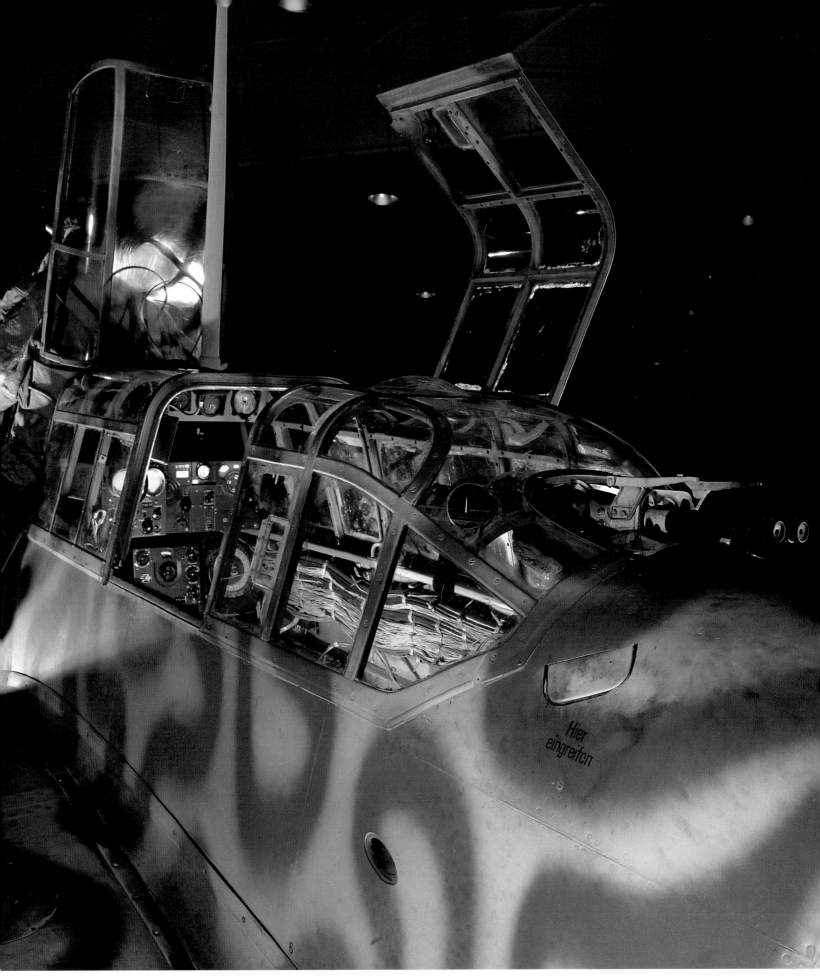

ABOVE: THE REAR GUNNER / RADAR RADIO OPERATOR'S POSITION IN THE BF 110G-4/R3 NIGHT FIGHTER. THE REAR GUNNER WAS ARMED WITH THE MG 81 TWIN 7.92. HIS JOB AT NIGHT WAS MORE PASSENGER THAN COMBATANT.
RAF MUSEUM, HENDON.

OPPOSITE TOP: AFTER THE BATTLE OF BRITAIN, THE MG 81 TWIN 7.92 MM WAS FITTED TO THE BF 110.
THIS EFFECTIVELY DOUBLED ITS FIREPOWER BUT THE BF 110 REMAINED VULNERABLE TO SINGLE-ENGINE FIGHTERS.

Junkers
Ju 88

THE JU 88 WAS THE TRUE BACKBONE OF THE LUFTWAFFE. PRODUCED IN GREATER NUMBERS THAN ALL OTHER GERMAN BOMBERS COMBINED, THE JU 88 PERFORMED EVERY TASK ASSIGNED TO IT WITH DISTINCTION.

I n many ways the Ju 88 was more versatile than the famous de Havilland Mosquito. Originally designed to serve as a medium-level bomber and dive bomber, the Ju 88 went on to serve as a long-range escort fighter, night fighter, intruder, pathfinder, tank buster, torpedo bomber, reconnaissance, transport and ground-attack aircraft. It truly was a jack-of-all-trades and Germany's finest medium bomber of the war.

As in most German bombers, the crew compartment was located at the front of the aircraft. Some would say that was typically Germanic, while British propaganda claimed the four crewman were

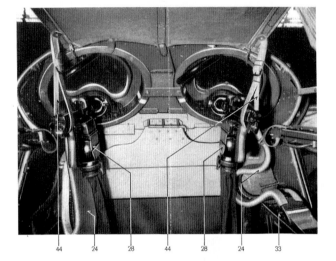

Abb. 13:   MG 81 im B-Stand am Zurriemen gezurrt, Gurtkasten eingesetzt

grouped together for morale reasons. In some ways, it was a cramped arrangement, but it also allowed the crew to operate more efficiently in an emergency where communication was critical. The pilot sat high on the left; the bomb aimer sat low on the right and operated the front gun mounted through the front windshield. Behind on the left was the flight engineer; he manned the upper rear armament. The radio operator, located on his right, manned the lower rear gun.

The Ju 88 was superior to both the Do 17 and He 111, but in 1939, production was slow. Early encounters with British fighters revealed how lightly armed the Ju 88 really was. By September 1940, no fewer than four separate 7.92-mm MG 15 machine guns were installed in the upper rear position. All were hand held and equipped with 75-round drum magazines that had to be changed after only three or four seconds' firing. This proved a major drawback in the heat of battle. There were as many as forty different armament schemes tried on the Ju 88, but, as with all bombers, none were truly effective.

In its most efficient role, the Ju 88 was a truly outstanding night fighter. The Ju 88C was the first night fighter version to go into mass production. Long endurance, tremendous performance, heavy armament and a wealth of electronic devices made the Ju 88 into an extremely formidable night fighter. Later versions were even more effective and, had they appeared earlier in the war, they would have posed a very serious threat to RAF Bomber Command's night-bombing campaign.

ABOVE: TAKEN FROM A GERMAN GUNNER'S MANUAL, THIS PICTURE SHOWS THE TWO REAR-MOUNTED MG 81 MACHINE GUNS. BECAUSE THE MG 15 WAS EXPENSIVE AND SLOW TO MAKE, THE MAUSER MG 81 BECAME STANDARD ISSUE ON ALL GERMAN BOMBERS.
RIGHT: THIS COCKPIT INTERIOR OF A JU 88A SHOWS THE PILOT AND THE FLEXIBLE, MOUNTED MG 15. THIS GUN WAS USUALLY MANNED BY THE BOMB AIMER.
INSET: THE JU 188 VERSION OF THE FAMOUS JUNKERS BOMBER WAS UP-GUNNED AND EQUIPPED WITH A 13-MM MG 131 ELECTRO-HYDRAULICALLY OPERATED TURRET AND AN AFT-FIRING 13-MM MACHINE GUN.

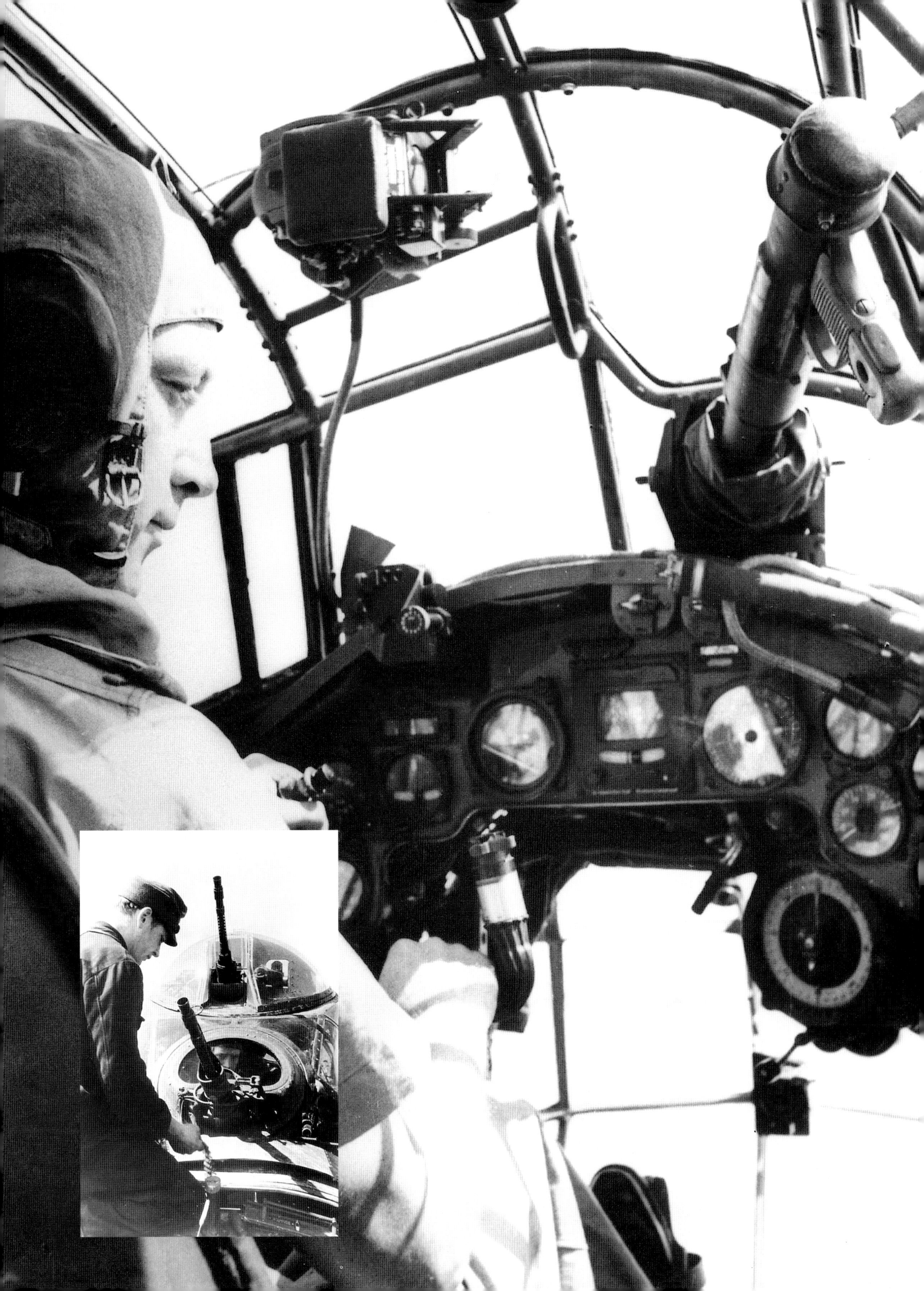

7
6
5
2
1
4
3
7412/10

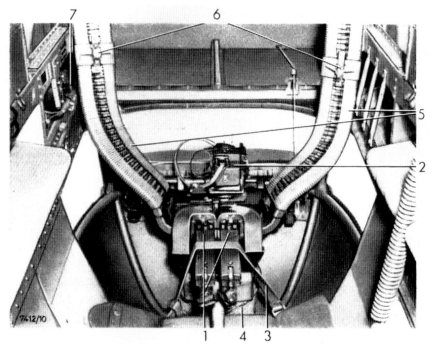

1 MG 81 Z 2 Revi 16 A 3 Zurrung 4 Leergutableitschlauch
5 Gurtzuführungsschläuche 6 Gurtbremsen 7 Drehknopf für Verdunkler

Abb. 12: C-Stand: MG 81 Z

Ju 88.
JUMO 211 J ENGINES
DIVE BRAKE (SHOWN FOLDED).
ENGINES JUMO 211 J INVERTED IN-LINE LIQUID-COOLED 1,300 H.P. AT 12,500 FT
EXHAUST SHIELD
BALLOON CABLE-CUTTER
POSITION FOR PRECISION BOMB-SIGHT.
KERRY LEE

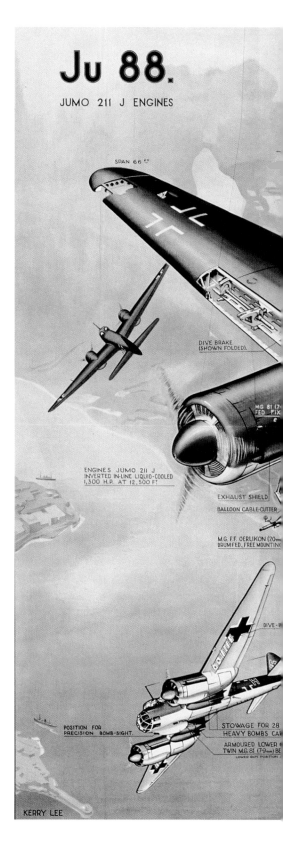

"Our crew shot down a total of fifty-nine enemy aircraft. In March 1945, we shot down nine bombers in one night — eight Lancasters and one Boeing B-17. I think the B-17 was a Pathfinder. Three of the nine I shot down using the rear-mounted MG 131 (13-mm) machine gun.

"We flew with four men in the Ju 88. There were two radio operators, one for the radar and one for the radios. I took care of the radios and manned the rear gun; I also handled the Naxos set, which homed in on the H2S radar carried by British bombers.

"We were on our way back to our base when I said to my pilot, 'I have targets in my Naxos.' We had never used this device before, so we decided to see if it would work. But there was a problem. After shooting down the six bombers, we were out of ammunition for the front MG 151 20-mm cannons. You couldn't reload the cannons in the Ju 88 the way you could in the Me 110. As we approached the H2S radar signature, we discovered three bombers flying very close together. We approached them from behind and underneath. It was decided that I should try to use the rear-mounted gun to bring the bombers down. As we flew under the first bomber, I had to direct my pilot because he didn't actually see the aircraft. We flew so close, I could actually see the crew and squadron identification number! I then took aim and fired between the engines and fuselage. We never fired at the cockpit. We always thought it was better to set the aircraft on fire — that way, the crew had a chance to escape. The next two bombers went down the same way. The last bomber shot down was a B-17, which, as it turned out, was equipped with the H2S radar.

"After we landed, we received a phone call from headquarters in Berlin. They were curious to find out how we shot down three bombers using the Naxos equipment."

Karl Johanssen, Me 110 / Ju 88 Radio Operator / Air Gunner

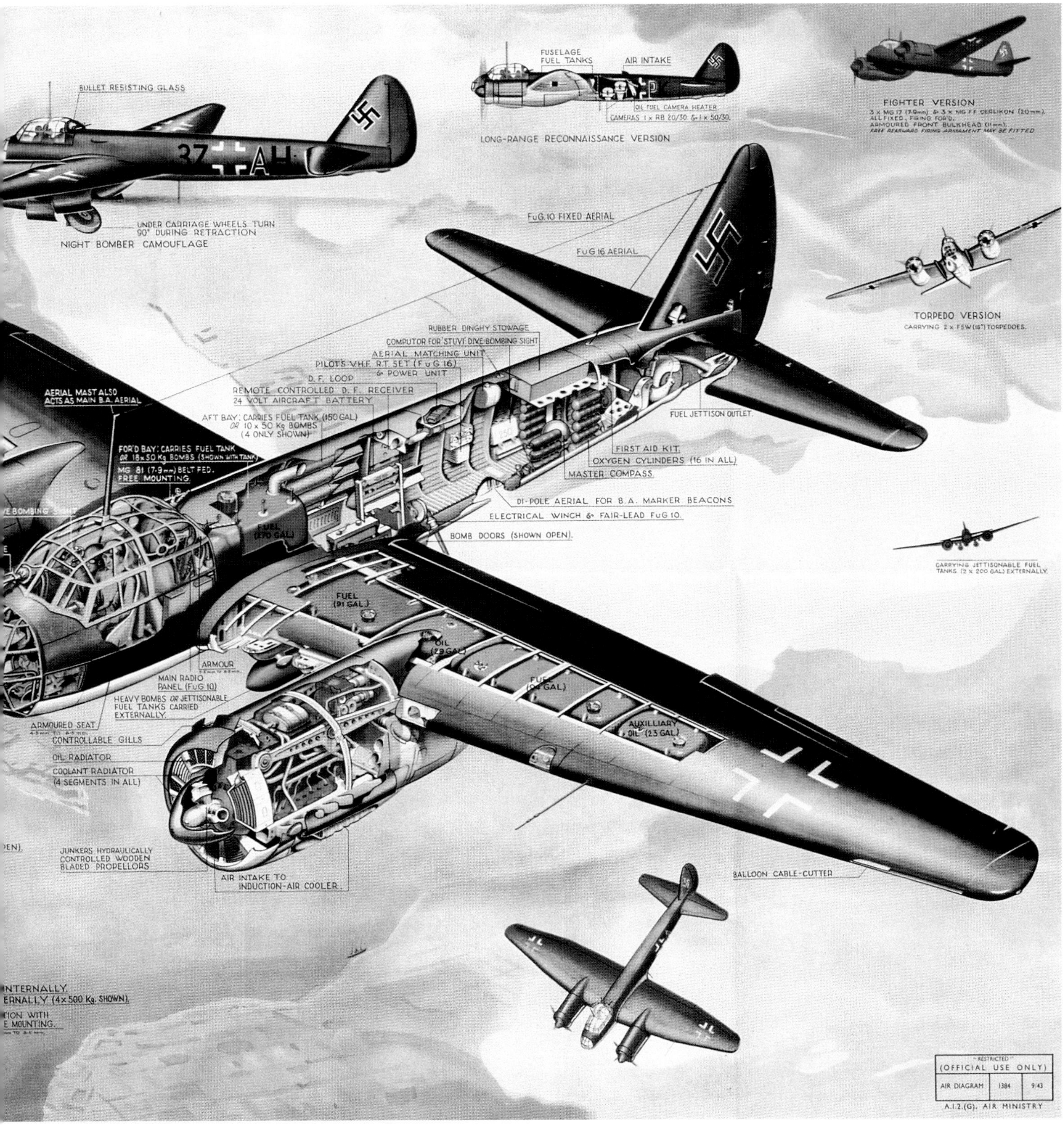

BULLET RESISTING GLASS
UNDER CARRIAGE WHEELS TURN 90° DURING RETRACTION
NIGHT BOMBER CAMOUFLAGE
FUSELAGE FUEL TANKS
AIR INTAKE
OIL FUEL CAMERA HEATER
CAMERAS 1 x RB 20/30 & 1 x 50/30
LONG-RANGE RECONNAISSANCE VERSION
FIGHTER VERSION
TORPEDO VERSION
FuG.10 FIXED AERIAL
FuG 16 AERIAL
RUBBER DINGHY STOWAGE
COMPUTOR FOR 'STUVI' DIVE-BOMBING SIGHT
AERIAL MATCHING UNIT
PILOT'S V.H.F. R.T. SET (FuG 16) & POWER UNIT
D.F. LOOP
REMOTE CONTROLLED D.F. RECEIVER
24 VOLT AIRCRAFT BATTERY
AERIAL MAST ALSO ACTS AS MAIN B.A. AERIAL
AFT BAY: CARRIES FUEL TANK (150 GAL) OR 10 x 50 Kg BOMBS (4 ONLY SHOWN)
FUEL JETTISON OUTLET
FIRST AID KIT
OXYGEN CYLINDERS (16 IN ALL)
MASTER COMPASS
DI-POLE AERIAL FOR B.A. MARKER BEACONS
ELECTRICAL WINCH & FAIR-LEAD FuG 10
BOMB DOORS (SHOWN OPEN)
FUEL (91 GAL)
ARMOUR
MAIN RADIO PANEL (FuG 10)
HEAVY BOMBS OR JETTISONABLE FUEL TANKS CARRIED EXTERNALLY
ARMOURED SEAT
CONTROLLABLE GILLS
OIL RADIATOR
COOLANT RADIATOR (4 SEGMENTS IN ALL)
AUXILLIARY OIL (23 GAL)
JUNKERS HYDRAULICALLY CONTROLLED WOODEN BLADED PROPELLORS
AIR INTAKE TO INDUCTION-AIR COOLER
BALLOON CABLE-CUTTER
"RESTRICTED"
(OFFICIAL USE ONLY)
AIR DIAGRAM 1384 9.43
A.I.2.(G). AIR MINISTRY

Above: The Ju 88 was one of the most versatile aircraft of the Second World War. An RAF Air diagram poster issued during WW II.

Left: The lower rear gun position in the Ju 88, armed with a pair of Mauser MG 81 7.92-mm machine guns.

HEINKEL
HE III

DURING THE BATTLE OF BRITAIN, THE HE III WAS ARMED WITH JUST THREE HAND-HELD
LIGHT-CALIBER MACHINE GUNS. THESE PROVED INEFFECTIVE AND SUBSEQUENT
VERSIONS CARRIED UP TO SEVEN MACHINE GUNS AND ONE 20-MM CANNON.

When Germany invaded Poland in September 1939, the Luftwaffe had 808 He 111s ready. It would be the most-used bomber in the German Air Force and bore the brunt of the opening bombing campaign in the West. Even though it was lightly armed, with just three 7.92-mm MG 15 machine guns, the He 111 enjoyed much success. Fighter opposition in Poland and Norway was minimal. The obvious weaknesses of the He 111 were not exposed. It was not until the Battle of Britain that the He 111 suffered heavy losses from the well-armed and fast Hawker Hurricane and Supermarine Spitfire. In response, the He 111 was up-gunned. Additional MG 15 machine guns were added, two machine guns on either side of the fuselage in the beam position and one extra gun added to the nose just to the right of the pilot's position. Later versions carried a 20-mm MG FF cannon in the nose. Even with the additional firepower, the He 111 was extremely vulnerable to fighter attacks. At the beginning of the Battle of Britain, the Luftwaffe had seventeen Gruppen equipped with the

He 111 — 500 aircraft, in all. In the course of the battle, 246 He 111 were lost.

All of the He 111's guns were hand held and of rifle caliber. The MG 15 gun had a good rate of fire of 1,000 rounds per minute, but it was not belt fed, so sustained fire could not be maintained. Each gun was fed from a 75-round saddle magazine, which translated into about four seconds of firing. In a prolonged engagement, the gunner was at a distinct disadvantage.

Throughout the war, the He 111's armament was added to continually. One of the final versions of the He 111 was equipped with an electric-powered dorsal turret armed with a single 13-mm MG 131 machine gun, twin 7.92-mm MG 81Z machine guns in the beam positions, one 13-mm gun in the ventral position, and one in the nose. At the beginning of the war, the He 111 was the Luftwaffe's principal bomber and one of the world's best. Despite continual improvements, it never regained its earlier prominence and ended the war in the transport role.

RIGHT: THE UNDERSIDE AND REAR OF THE HE III WAS SOMEWHAT PROTECTED BY SINGLE GUN IN THE BELLY GONDOLA.
THIS PHOTO SHOWS A VENTRAL GONDOLA GUNNER LYING IN HIS POSITION.
BELOW: A GROUP OF HE III BOMBERS FLYING OVER A GERMAN PORT ON THEIR WAY TO A TARGET IN BRITAIN.

"A couple of times I found myself in rather large German formations, and I can remember the engine noise over everything else. They were all German engines! I could hear them over my own engine in the Hurricane! Of course, their air gunners were firing at me, but it was a question whether they hit me or not. The one thing you have to remember is, the moment you opened fire you lost 14 miles an hour of air speed. The recoil from eight machine guns firing at once actually slowed you down!

"We tried to attack from above and behind, but it didn't always work out like that. Normally we were just catching up with the German bombers. There wasn't time to hang about. You had a good old squirt and then you left, because if you hung about, you were killed. The German air gunners during the Battle of Britain were pretty well disciplined and they seemed to hold off until you got close enough. The Germans used a flat formation. They tended to bunch up to get maximum firepower from their rear gunners."

DENNIS DAVID, HURRICANE PILOT

BRITISH REPORT T.C. 10 GERMAN BOMBER FORMATIONS 1939–1940

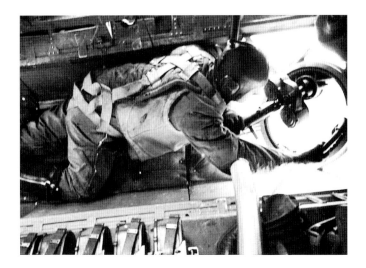

TACTICS OF GERMAN BOMBERS AGAINST BRITISH FIGHTERS

21. The large German bomber formations did not usually adopt any evasive tactics and rely for their protection against fighters on:

(i) Fighter escorts
(ii) Close formation flying for mutual fire support.

Fine weather being a primary requirement for operating mass formations, use of cloud cover is precluded.

22. When smaller bomber formations have been encountered by our fighters, or when the large formations have been broken up, the bombers, either individually or in sections, have adopted one of the following evasive tactics.

(i) Sought the nearest cloud cover
(ii) Descended to ground or sea level at high speed and making full throttle, low, get away, accompanied by "jinking."
(iii) Used diving brakes, when fitted, to enable a steep dive, combined with turns, to be executed, thus causing attacking fighters to overshoot.
(iv) Simulating damage by emitting smoke from an apparatus fitted either in the tail or behind the engine nacelles.

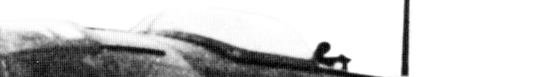

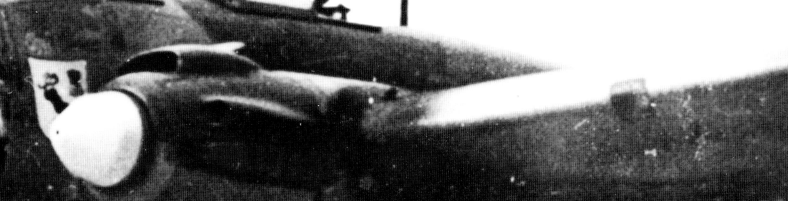

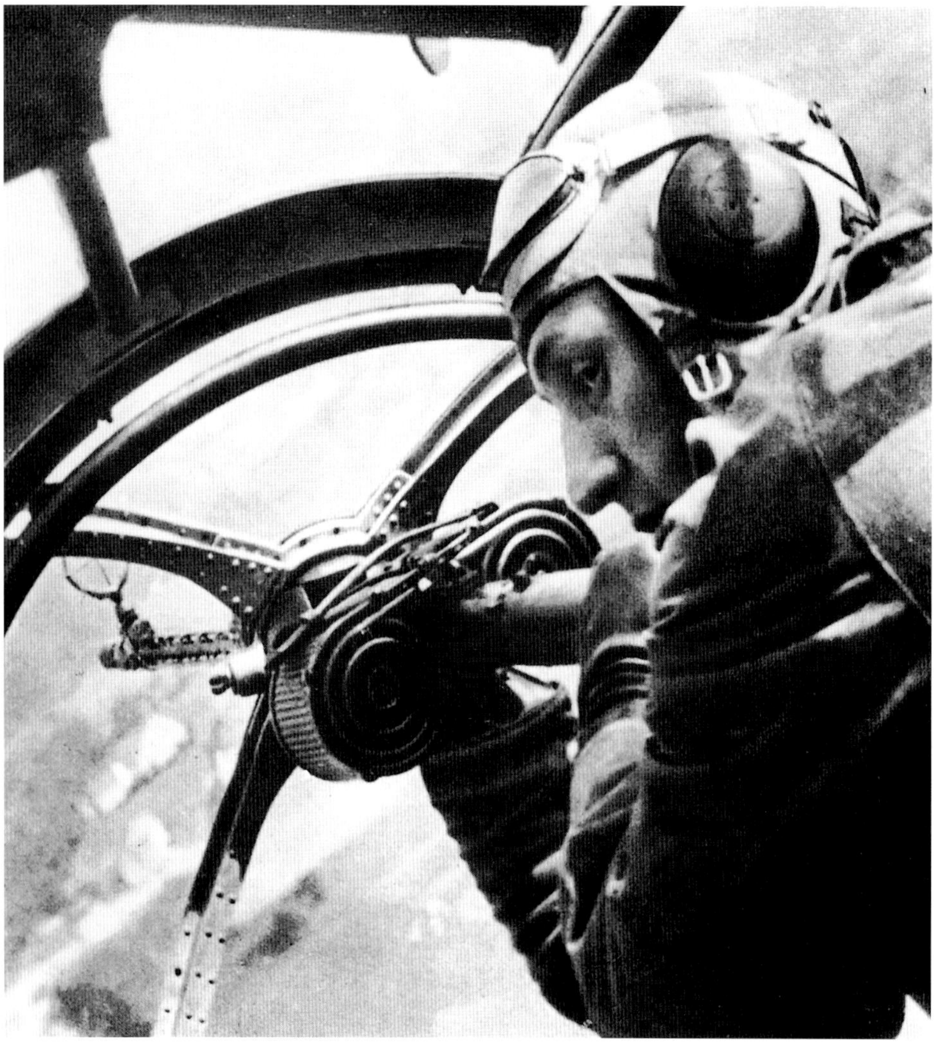

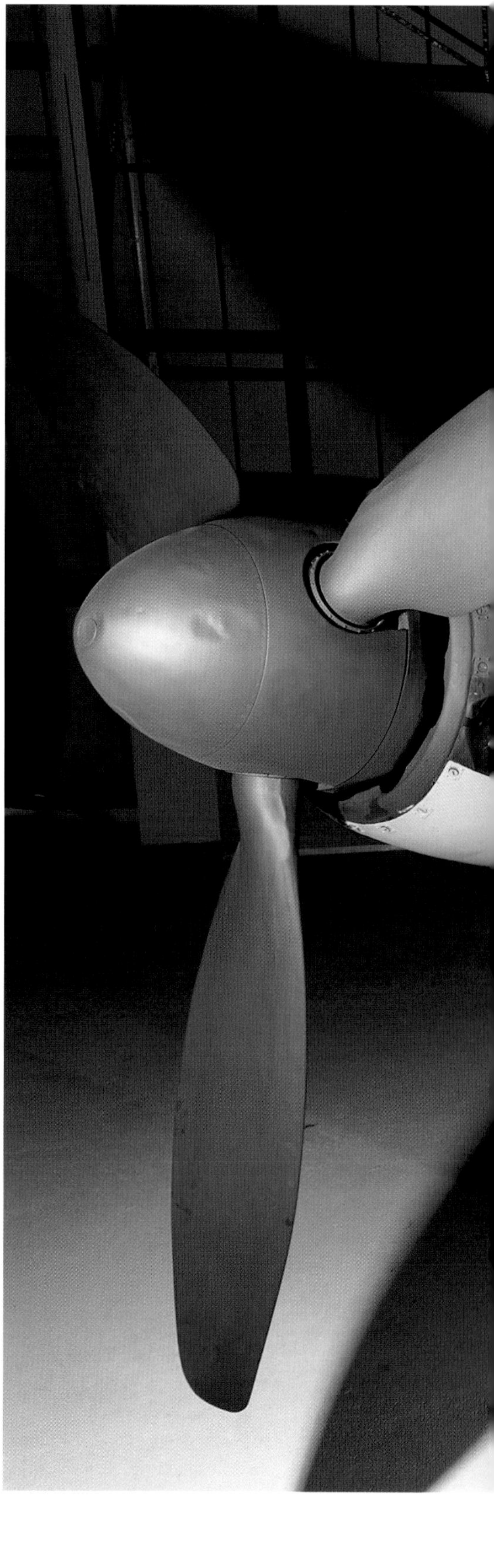

A. Kurzbeschreibung

Der B-Stand befindet sich an der Rumpfdecke zwischen den Spanten 8—12a.

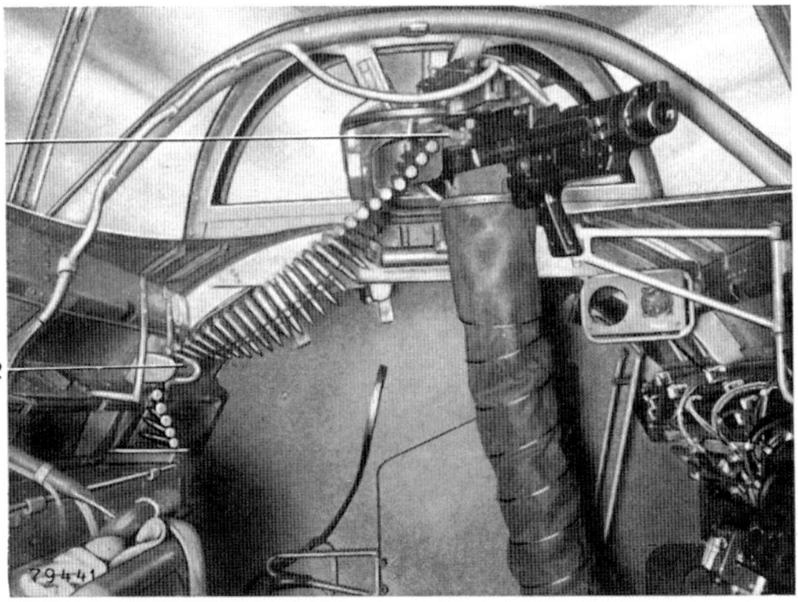
1
2
79441

1 Gurtzuführung 2 Gurtumlenkrolle

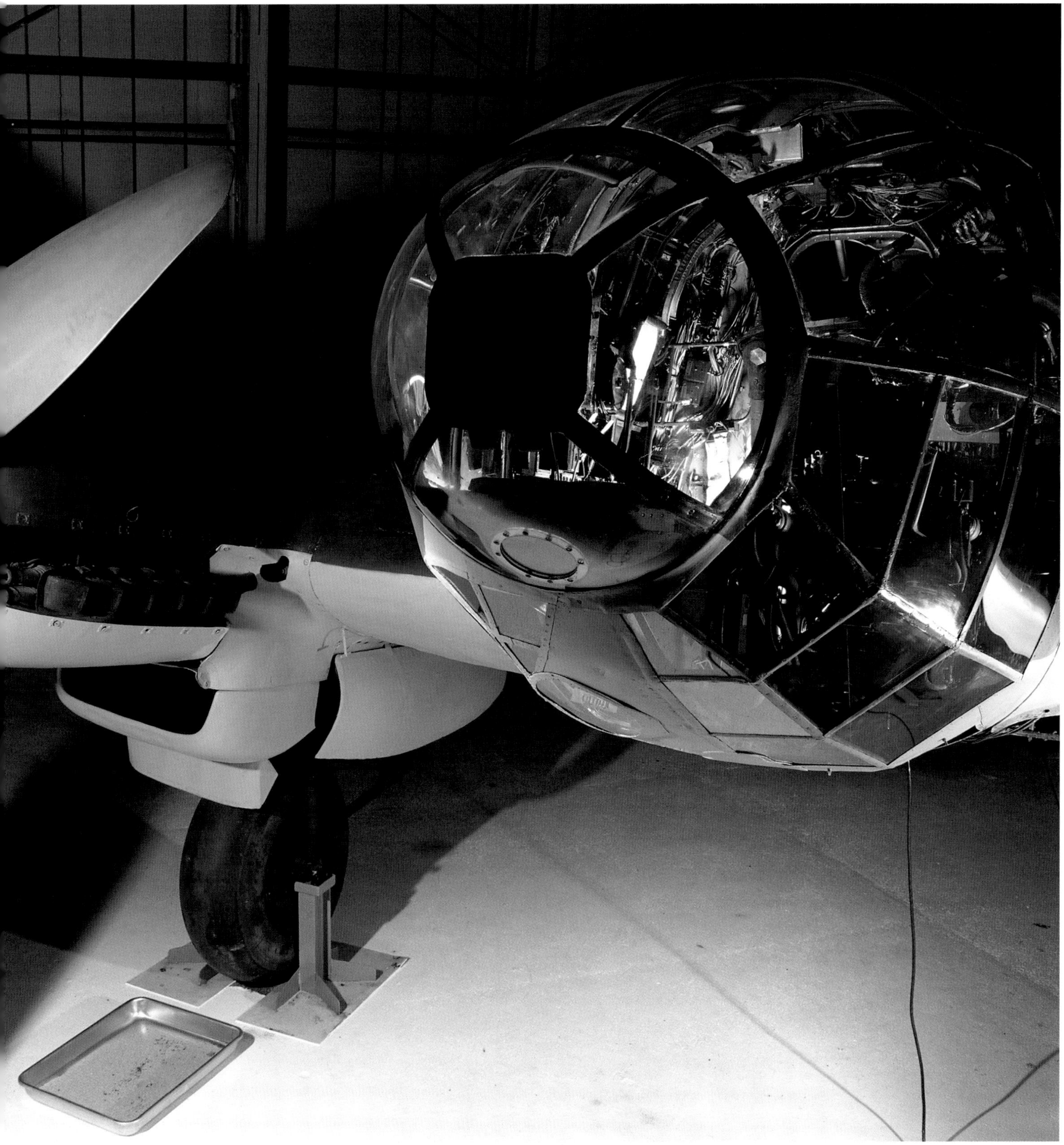

Above: The extensively glazed nose of a He 111H-23 transport.
RAF Museum, Hendon.

Top left: Nose gunner in an He 111 searches for ground targets.

Bottom left: The He 111H-11 was the first to have a fully enclosed dorsal gun position and was armed with a single MG 131 13-mm machine gun.

Messerschmitt

Even with its advanced rear defensive armament, the Me 410 was no match for marauding Spitfires, Mustangs and Thunderbolts.

As the USAAF daylight bombing campaign increased, so did the German resistance. More and more fighter units were transferred from other fronts to bolster the home defense. American air gunners not only faced heavily armed single engine fighters, but they also had to deal with twin-engine fighters like the Bf 110 and Ju 88. These heavily armed *Zerstörer* (destroyer) fighters carried rockets, multiple cannon packs and, in some cases, a single 37-mm or 50-mm cannon. They were extremely effective. By mid-1943, air gunners began to encounter a new fighter — the Me 410.

The Messerschmitt Me 410 was originally designed to replace the Bf 110, but early development problems and poor performance of the original Me 210 delayed the Me 410's introduction. Even when it was ready for combat, the Me 410 was not used in the long-range strategic fighter role. Germany was now on the defensive and a large number of Me 410s were committed to the defense of the homeland. Although its performance was good, the Me 410 was no match against a Spitfire or Mustang. Against unescorted bombers, the Me 410 was extremely

effective. The Me 410 could be armed with up to six 20-mm cannons, two 7.92-mm machine guns, and four Werfergranate (Wgr) 210-mm air-to-air rocket-propelled motor shells mounted under the wings. This weapon was an adaptation of the standard 210-mm infantry motor shell. Its flight speed was around 320 meters (1,050 ft) per second and it was usually fired from a distance of 1,200 meters (1,320 yards). Terribly inaccurate, the Wgr air-to-air rocket was not designed to shoot down individual aircraft; its sole purpose was to break up the tight American formations and scatter the bombers. Once this was accomplished, the helpless bombers were easy prey for marauding fighters.

A number of Me 410s also carried the 50-mm BK 5 cannon. This was a stand-off weapon designed to let the Me 410 fire from outside of the protective range of the .50-caliber machine gun. A single hit by a 50-mm shell was usually fatal to a large four-engine bomber. In the end, the slow rate of fire and the difficulty of hitting a moving target with such a large weapon proved impractical.

By the summer of 1943, American bombing raids had increased in their size and effectiveness. The Luftwaffe was forced to pull back more and more fighters from other fronts. The Bf 110 and Me 410 were ideally suited to attack the heavy bombers. On March 16, 1944, a mixed force of Bf 110 and Me 410s attacked a force of 500 B-17s and 200 B-24s. Using rockets and cannons, the *Zerstörer* units shot down eighteen heavy bombers. In April, the 8th Air Force attacked again with 900 B-17s and B-24s. A mixed formation of Ju 88s and Me 410s destroyed a total of twenty-five bombers, but this rate of success would not last. By mid-July 1944, Allied fighter protection was now measured in the hundreds and there was no longer a safe place in German airspace for the slower Me 410 and Bf 110. After heavy losses, the Me 410 was withdrawn from service.

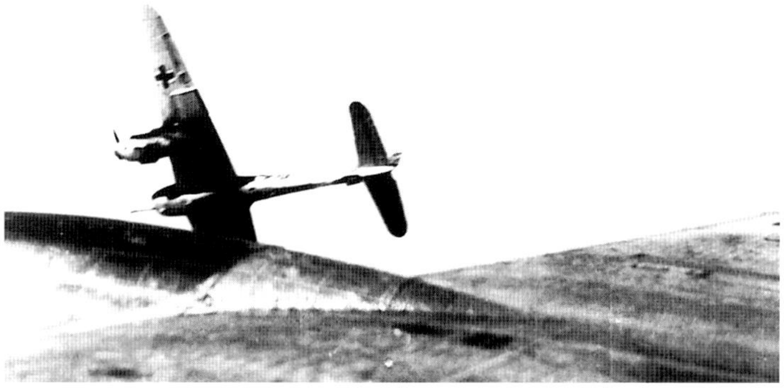

The Me 410 was a fast, heavily armed fighter, and one of its most unique features was the remotely controlled FDL 131 barbette gun turret system. This advanced feature consisted of a large, transversely mounted drum located in the fuselage just aft of the wing. This drum was rotated up and down by an electric motor. On the drum's left and right sides were mounted single 13-mm MG 131 machine guns. Each gun had 450 rounds of ammunition. These guns could traverse through 70 degrees above and below the horizon, through 40 degrees of azimuth and out to the 90-degree beam position. The observer/gunner who faced aft operated the entire FDL 131 with one pistol grip and two optical sights. The barbette system offered excellent firepower almost over the entire rear hemisphere with very little drag.

In July 1944, Captain James Morris, the leading P-38 ace of the 20th Fighter Group, was shot down by the side-mounted 13-mm guns of an Me 410 that he was attacking. As advanced as this remote-control turret system was, it did not stop American Mustangs, Thunderbolts and Lightnings from hacking the Me 410 from the skies over Germany.

Above: A remarkable photograph of an Me 410 armed with a single 50-mm BK 5 cannon breaking away after making a firing pass at a 388th Bomb Group B-17.

Below: A gunner in a Me 210 prepares his guns for action. The Me 210 proved up to the job, but like its successor, the Me 410 was dogged by an appalling reputation.

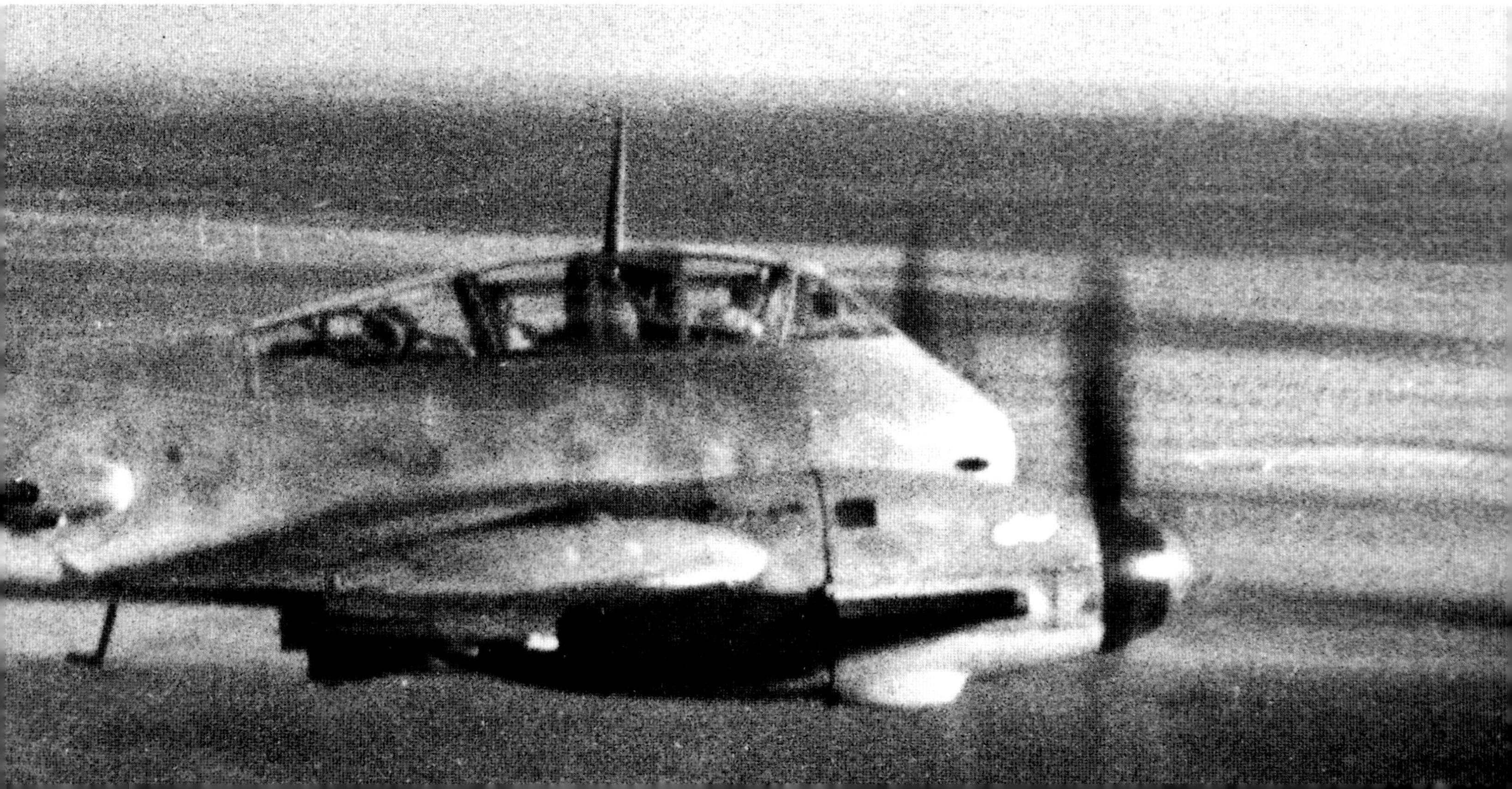

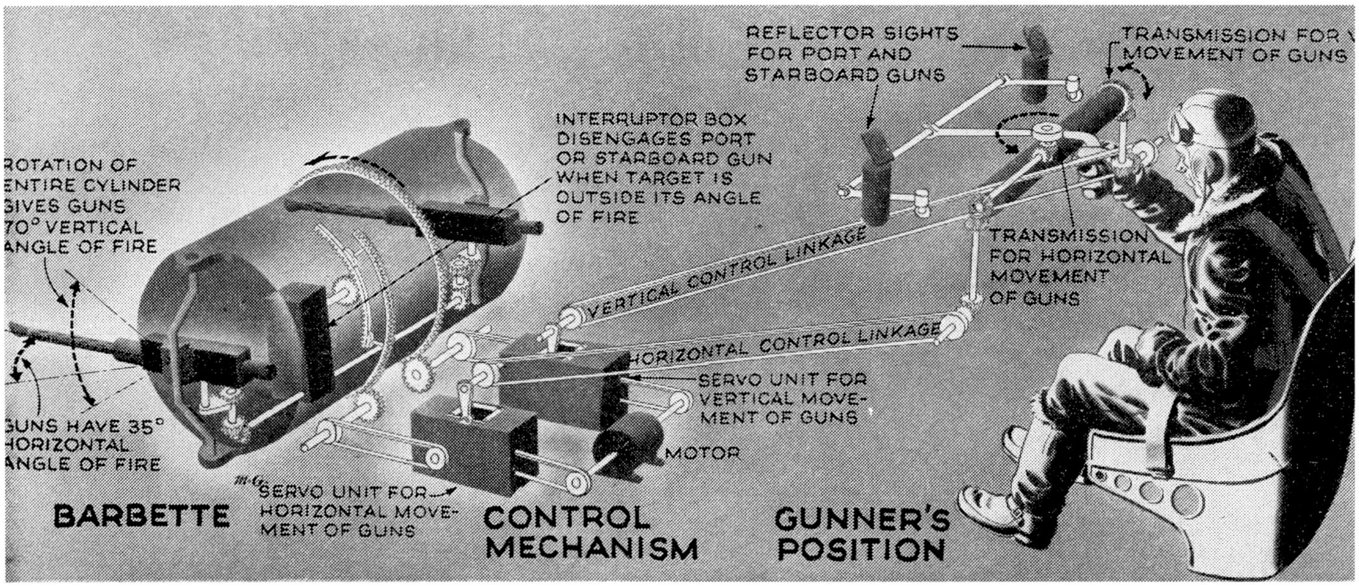
REFLECTOR SIGHTS FOR PORT AND STARBOARD GUNS
TRANSMISSION FOR VERTICAL MOVEMENT OF GUNS
INTERRUPTOR BOX DISENGAGES PORT OR STARBOARD GUN WHEN TARGET IS OUTSIDE ITS ANGLE OF FIRE
ROTATION OF ENTIRE CYLINDER GIVES GUNS 70° VERTICAL ANGLE OF FIRE
VERTICAL CONTROL LINKAGE
TRANSMISSION FOR HORIZONTAL MOVEMENT OF GUNS
HORIZONTAL CONTROL LINKAGE
SERVO UNIT FOR VERTICAL MOVEMENT OF GUNS
GUNS HAVE 35° HORIZONTAL ANGLE OF FIRE
MOTOR
SERVO UNIT FOR HORIZONTAL MOVEMENT OF GUNS
BARBETTE
CONTROL MECHANISM
GUNNER'S POSITION

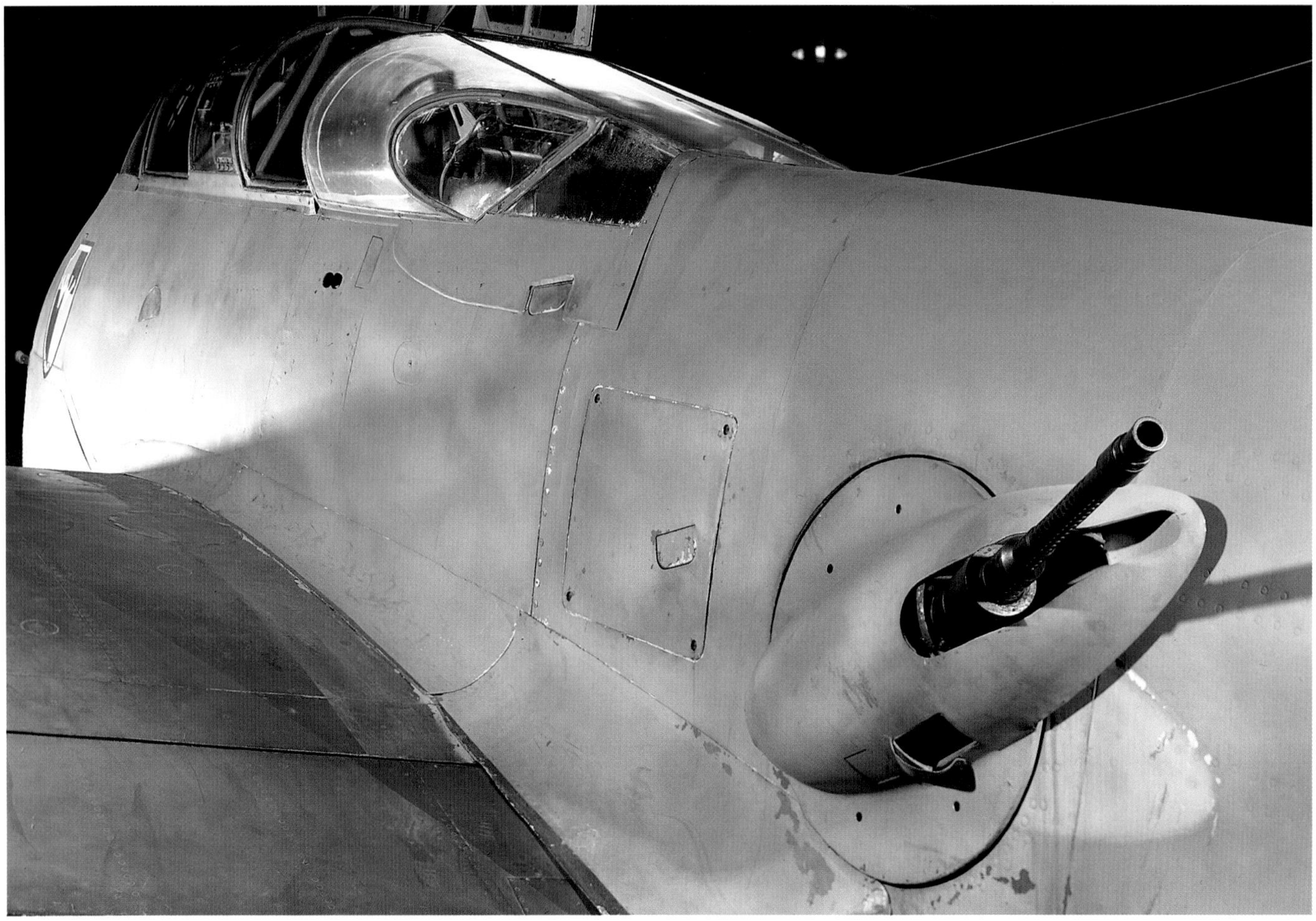

Top: A schematic diagram showing the method of operation of the FDL 131 remotely controlled gun system.

Above: Close-up of the side-mounted barbette. The MG 131 machine gun was belt-fed and fired up to 960 rounds a minute and weighed only 40 pounds. Note the open spent cartridge chute beneath the weapon.
Aerospace Museum, Cosford.

Right: The radio operator / gunner's position showing the centrally mounted pistol grip. The FDL remotely controlled guns system offered improved fields of fire in both the upper and lower hemisphere.
Aerospace Museum, Cosford.

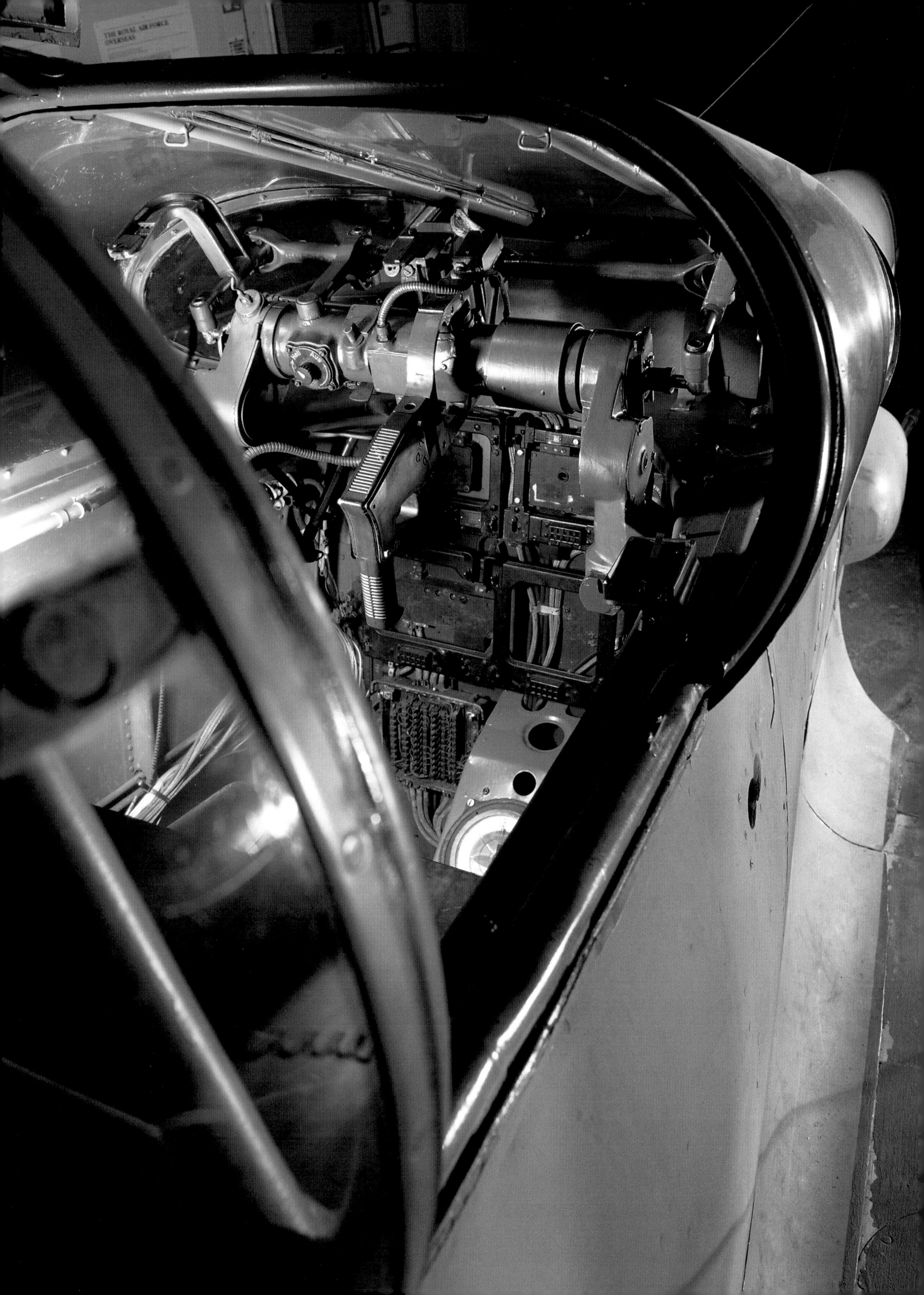
Hell
Aus

Savoia-Marchetti S.M. 79
Sparviero

The backbone of the Regia Aeronatica, the S.M. 79 was capable of suffering enormous battle damage and returning its crews safely home.

During the Second World War, the Italian Armed Forces were described as woeful and inadequate by Allied propoganda. Many believed that the Regia Aeronautica's combat aircraft were inferior to those of other European and North American air forces. This belief had little foundation. Italy captured numerous international aviation records before the war and its air force was equipped with some very sound, modern military aircraft. One such aircraft was the handsome Savoia-Marchetti S.M. 79. While the aircraft produced by the Italians during the war were as good or better than some Allied types, the inability of the Italian aircraft industry to keep pace with combat attrition proved the Regia Aeronautica's Achilles heel.

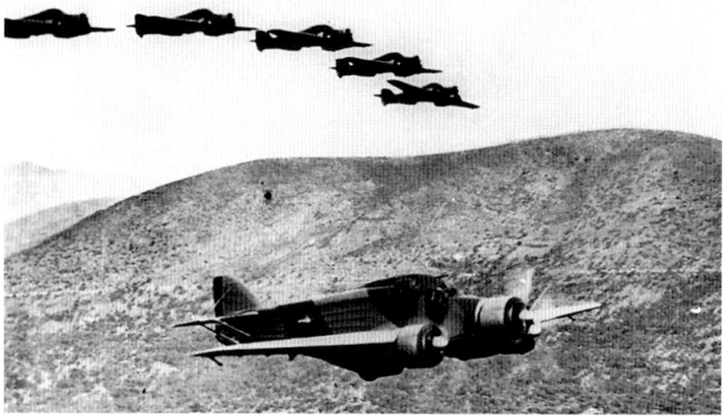

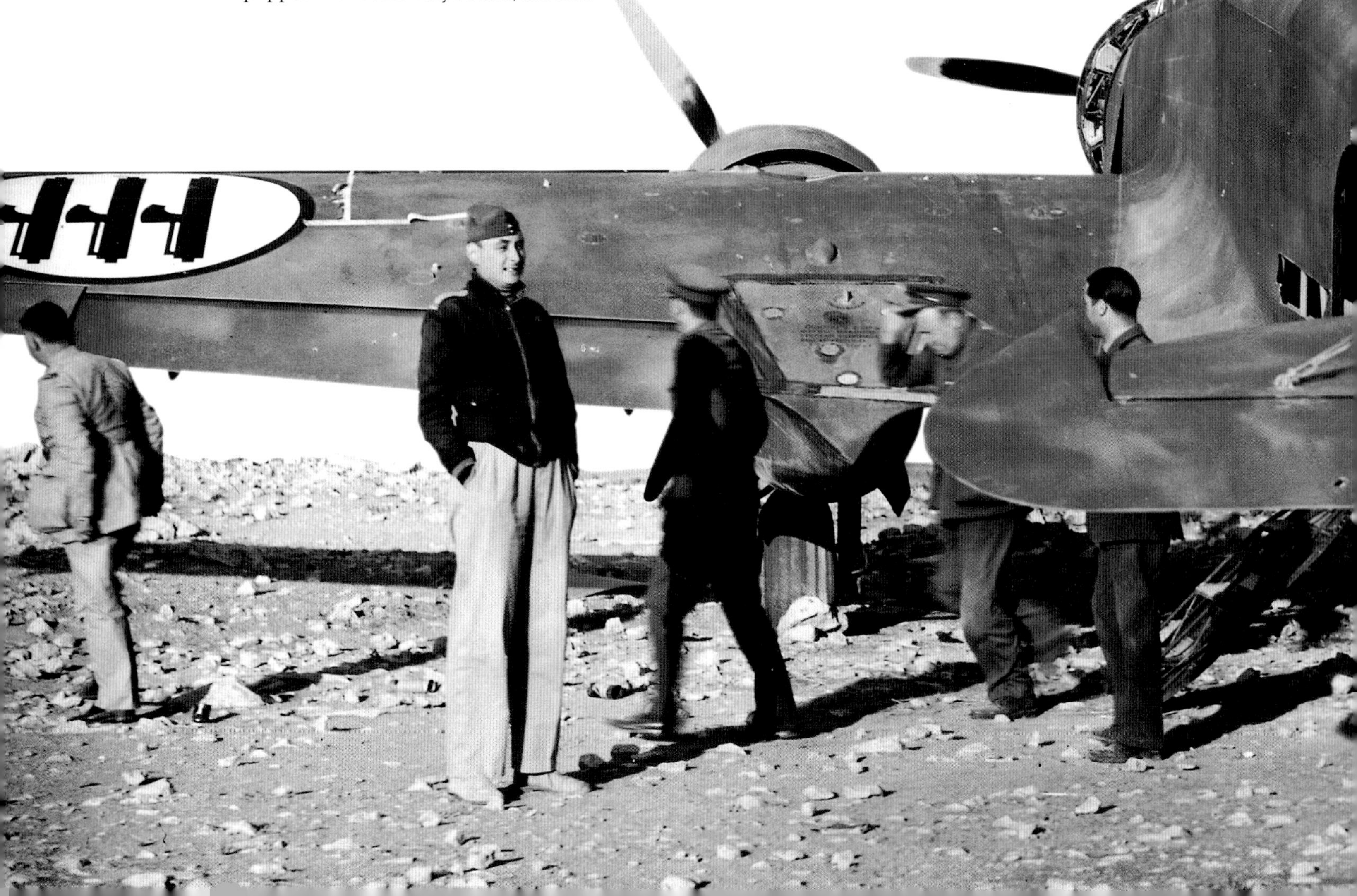

The S.M. 79 began as a eight-seat passenger high-speed commercial aircraft. The first prototype appeared in late 1934 and possessed exceptionally sleek contours. Within a year, the S.M. 79 established 1,000- and 2,000-km closed-circuit speed records with a payload of 2,000 pounds. Average speeds were 242 mph and 236 mph. The following year the S.M. 79 received more powerful engines and bettered its previous records by carrying a 4,000-lb payload over 1,000 km at an average speed of 266 mph. It was now clearly obvious that the S.M. 79 was better suited for military operations than to commercial passenger service. As a bomber, the S.M. 79 did not differ much from its civilian predecessor. The big difference was the streamlined fairing added above the flight deck to house the upper gun (fixed forward-firing) and the dorsal gunner's defensive position.

The S.M. 79 was the backbone of the Regia Aeronautica and equipped most of its bomber squadrons. It was extremely efficient and was perhaps the best and most successful land-based torpedo bomber of the war. The S.M. 79 was well liked by its crews, and to the Italian nation, the Sparviero was seen in the same light as the British Spitfire and Japanese Zero were by their countries. It was the pride of a nation and was flown by Italy's most famed wartime pilots.

The unique feature that separated the S.M. 79 from other bombers was its three-engine layout. While other air forces believed in the need for nose armament and adopted the twin-engine formula, the S.M. 79 bucked the trend. Because of the speed of the S.M. 79, the Italians believed that frontal attacks by fighters were unlikely. In any case, the S.M. 79 was equipped with a forward-firing Breda SAFAT 12.7-mm machine gun with 350 rounds,

This S.M 79 landed on the edge of Sidi Barrani aerodrome after being forced down by British fighters.
Left: A flight of S.M. 79s.

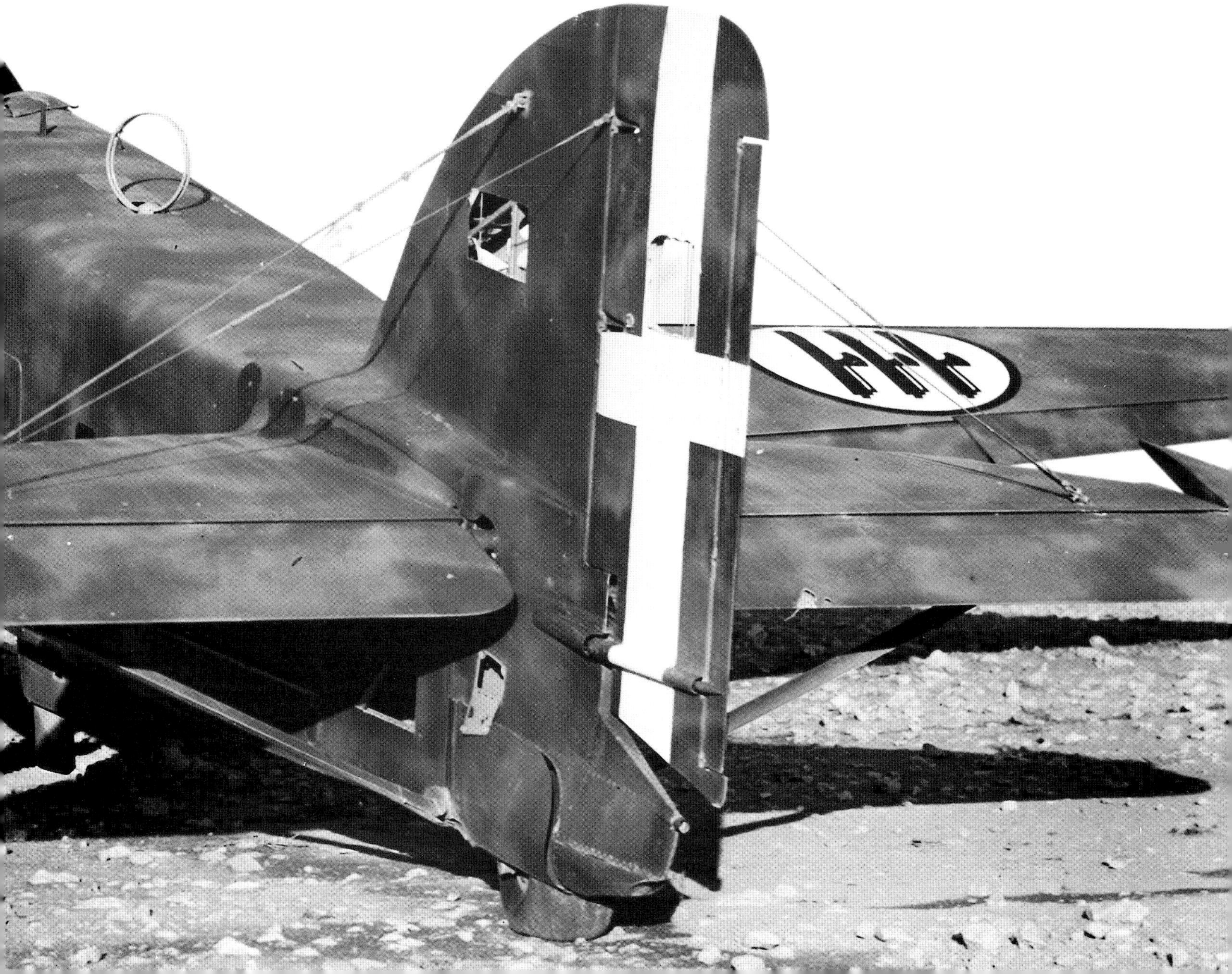

along with two flexible guns (500 rounds each) of the same caliber in the dorsal and ventral position, and one 7.7-mm (.303) Lewis gun in the waist position. The waist gun was mounted on a lateral bar and could fire out of either waist window. The ventral gondola was manned by the bombardier. The bombs in the S.M. 79 were stored vertically. A typical load would consist of two 1,100-lb. bombs, five 550-lb. bombs, twelve 220-lb. bombs, or two 450-mm torpedoes slung underneath the fuselage.

When war broke out in 1939, the Regia Aeronautica was equipped with 975 bombers, of which 594 were S.M. 79s. Much of the S.M. 79's success was gained as a torpedo bomber. During the war, the S.M. 79 achieved many victories including the sinking the carrier HMS *Eagle*, the destroyers *Jaguar, Legion, Southwall, Husky, Kujavik II*, as well as a number of other smaller naval vessels and cargo ships.

Although the S.M. 79 gained fame as a torpedo bomber, it was also used for strategic reconnaissance, level bombing, and close air support. In it most unusual role, the S.M. 79 was used as remotely controlled guided bomb. Packed with explosives and equipped with a radio-controlled guidance system, a pilotless S.M. 79 was prepared for a one-way mission. In August 1943, as the British Fleet lay off the coast of Algeria, the single radio-controlled S.M. 79 took off but never made it to the target. A faulty radio circuit caused the aircraft to veer off course and crash in the Klenchela mountains in Algeria.

The total production run of the S.M. 79 was just 1,200 aircraft. Although its numbers were small, it was undoubtedly one of the most successful medium bombers produced by the Italian aircraft industry. Its handling characteristics were excellent and crews favored it above all others.

Left: Interior of the upper dorsal position in the S.M. 79.
Museo Storico del' Aeronautica Militaire, Italy.

Below: The unique armadillo-type shell covering the upper dorsal gunner's position is evident. When combat was imminent, the gunner would push the sections back to improve his vision and field of fire.
Museo Storico del' Aeronautica Militaire, Italy.

Right: An Italian gunner in an unknown bomber scans the sky for enemy fighters.

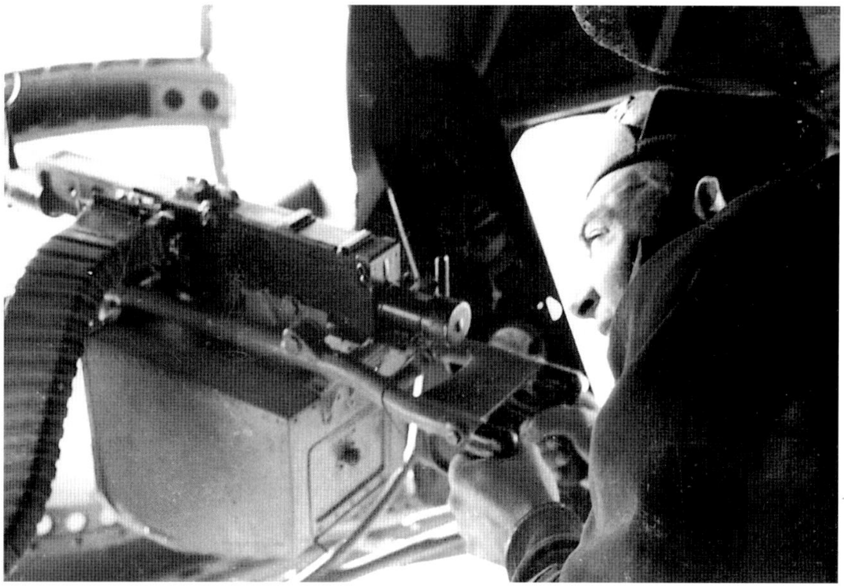

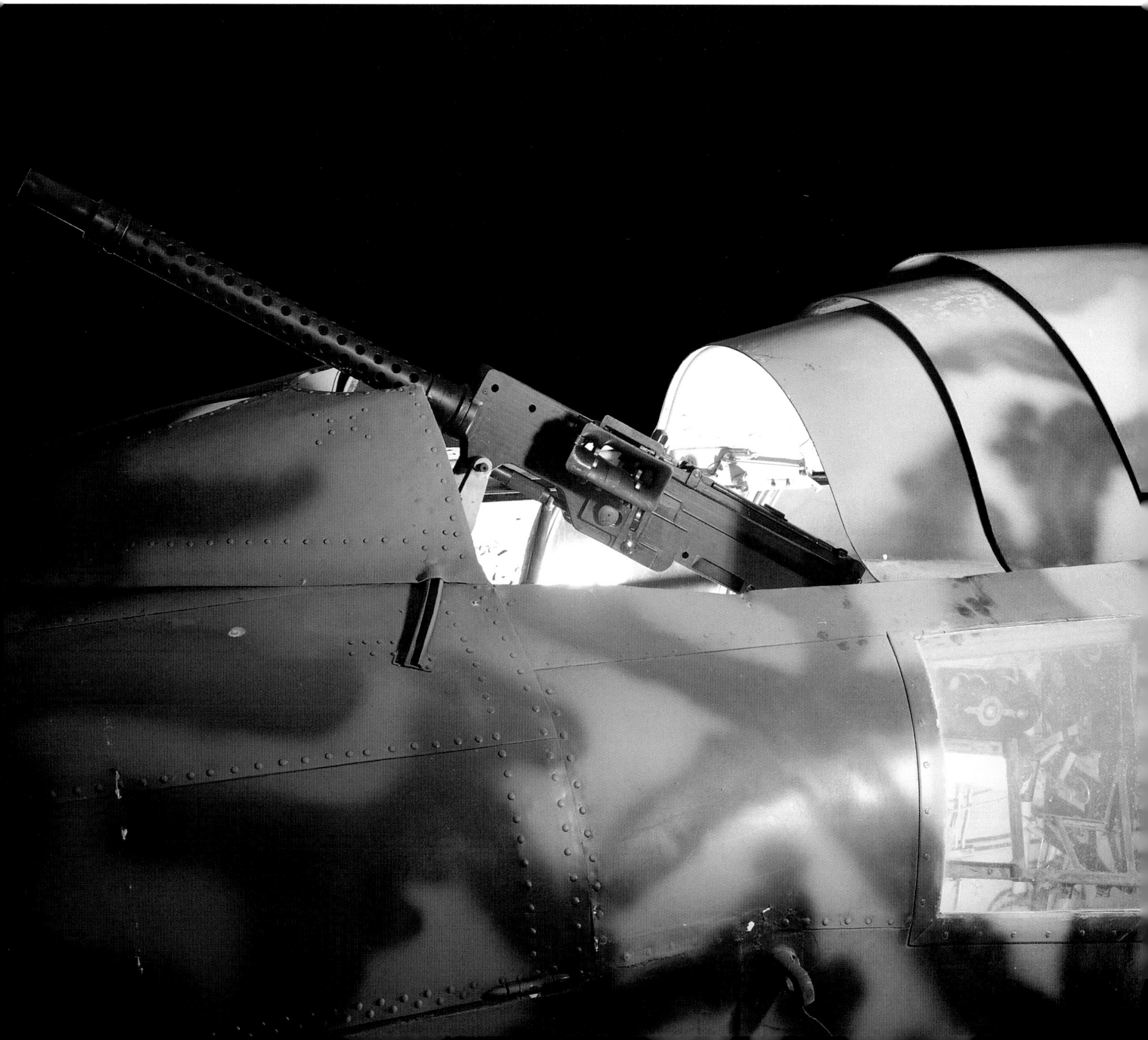

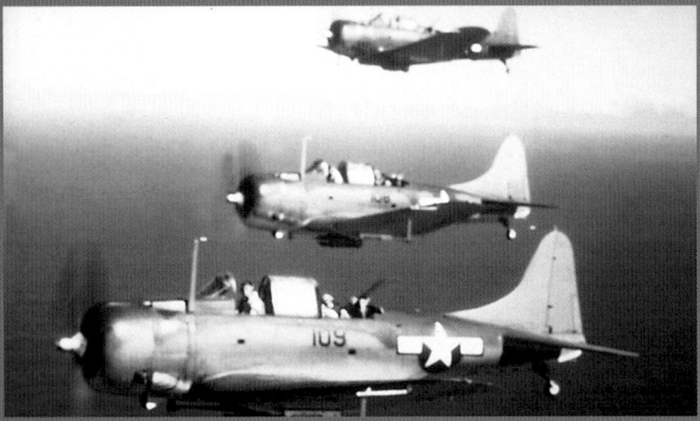

109

Douglas SBD Dauntless

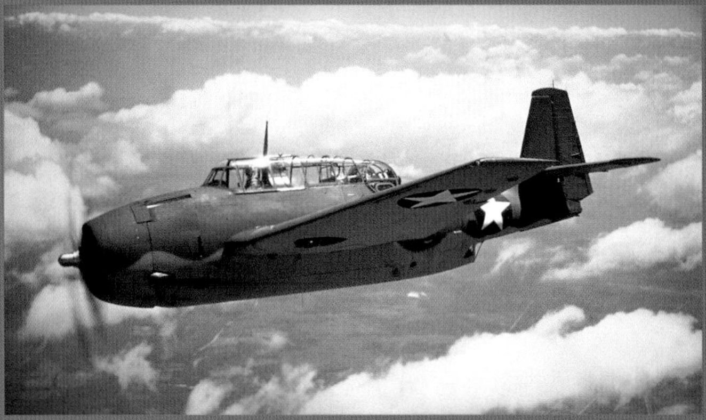

Grumman TBF Avenger

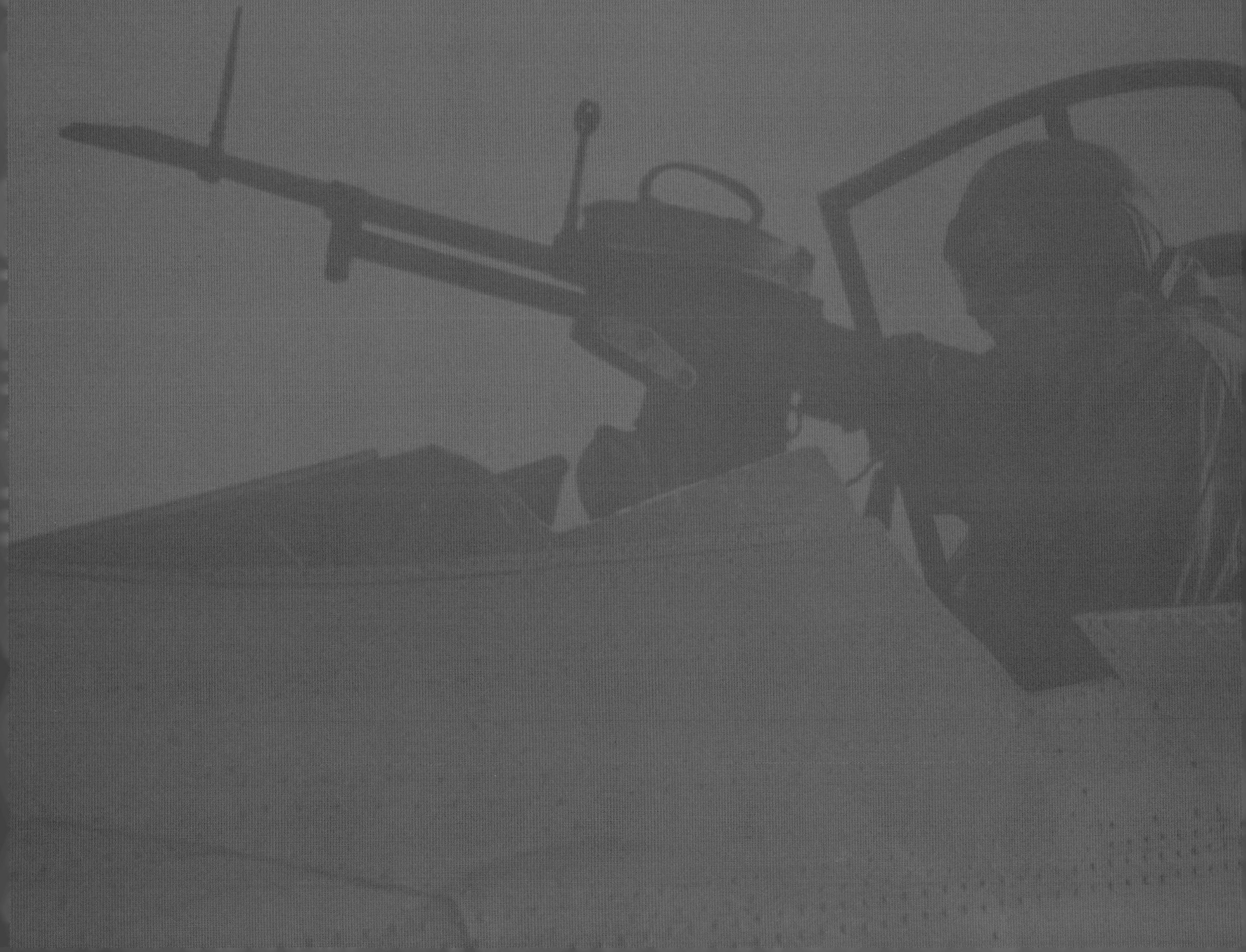

SINGLE ENGINE

◉ FAIREY BATTLE

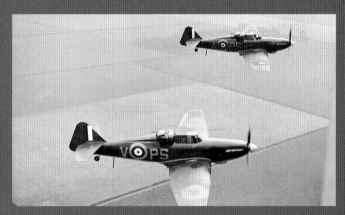

◉ BOULTON PAUL DEFIANT

☆ ILYUSHIN IL-2 SHTURMOVIK

Douglas SBD
DAUNTLESS

NOT ONLY DID THE SBD RADIOMAN/GUNNER HAVE TO DEFEND AGAINST ENEMY FIGHTERS, HE ALSO HAD
TO BE PROFICIENT IN RADIO VOICE AND MORSE CODE COMMUNICATIONS.

The Dauntless dive-bomber was in many ways similar to the Fairey Swordfish. Considered obsolete when the war started, it would go on to prove itself to be a crucial and effective weapon. The Dauntless served in six air arms — the US Navy, US Marines, US Army Air Force, French Air Force and the Royal New Zealand Air Force. Its been said that the battle for Guadalcanal could have been lost if the SB2C Helldiver had replaced the Dauntless before the war started. The crude, dusty conditions on Guadalcanal would have crippled the more sophisticated Helldiver. Its sortie rates would have been much lower, but the "obsolete" Dauntless proved to be reliable and easily maintained.

Like all Navy carrier-borne bombers and torpedo planes, the Dauntless was equipped with a rear gun. The radio operator / gunner manned a pair of drum-fed .3-in (7.62-mm) Browning machine guns; he was also responsible for both voice and Morse code radio communication.

A typical gunner trained for a year or more in communications, electronics, gunnery, and operational training before reaching the combat zone.

In 1942, the Dauntless played a major part in the victories during the Battle of the Coral Sea, Midway and Guadacanal. It also racked up an impressive record of air-to-air kills. The Dauntless was credited with ninety-four aerial victories and most were evenly divided between pilot and gunner. The most famous air-to-air victory claimed by Dauntless rear gunners was over Japanese ace Saburo Sakai. On August 7, 1942, while flying over Guadalcanal, Sakai spotted a formation of what he thought were F4F Wildcat fighters. The Dauntless gunners saw the Japanese fighters approaching and from long range they opened up. Sakai's Zero was hit and he was severely wounded. His aircraft dropped from 8,000 feet to sea level. He managed to fly 550 miles back to his base in Rabual.

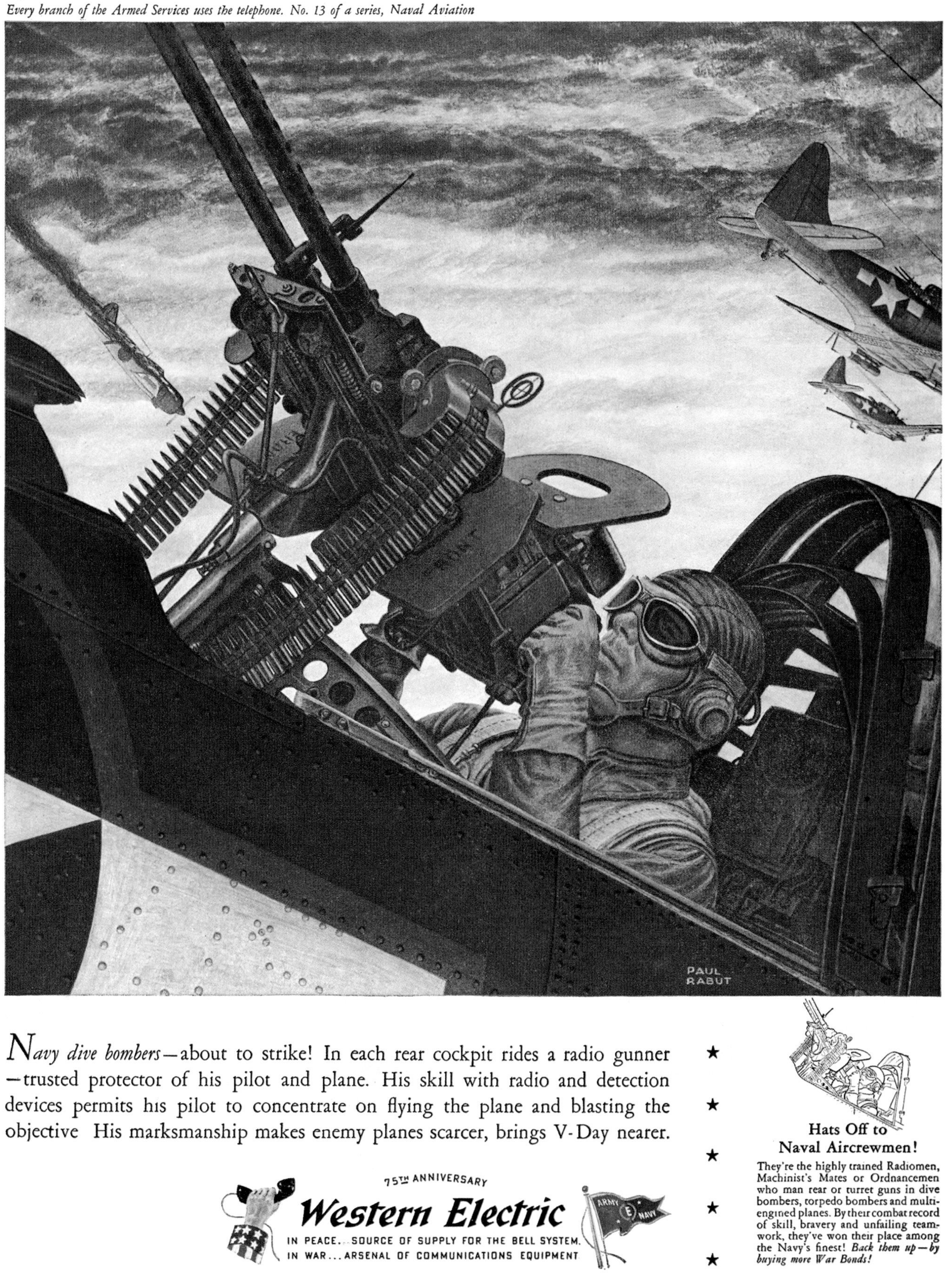

Every branch of the Armed Services uses the telephone. No. 13 of a series, Naval Aviation
PAUL RABUT
Navy dive bombers—about to strike! In each rear cockpit rides a radio gunner—trusted protector of his pilot and plane. His skill with radio and detection devices permits his pilot to concentrate on flying the plane and blasting the objective His marksmanship makes enemy planes scarcer, brings V-Day nearer.
Hats Off to Naval Aircrewmen!
They're the highly trained Radiomen, Machinist's Mates or Ordnancemen who man rear or turret guns in dive bombers, torpedo bombers and multi-engined planes. By their combat record of skill, bravery and unfailing team-work, they've won their place among the Navy's finest! Back them up—by buying more War Bonds!
75TH ANNIVERSARY
Western Electric
IN PEACE... SOURCE OF SUPPLY FOR THE BELL SYSTEM.
IN WAR... ARSENAL OF COMMUNICATIONS EQUIPMENT
ARMY E NAVY

LEFT: SBD-4 SKIMS OVER LOW-HANGING CLOUD WITH REAR GUNNER AT THE READY.
ABOVE: THIS WONDERFUL COMPOSITION OF A DAUNTLESS RADIO/GUNNER EVOKES THE GLAMOR ASSOCIATED WITH AIRCREW STATUS.

LIFE RAFT &
EMERGENCY
RATIONS

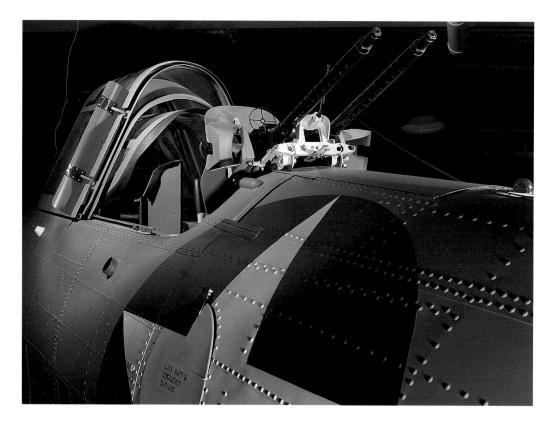

"I found the gunner's position in the SBD quite comfortable. It was also very windy. If you stayed behind the canopy cover, it was fine, but if you were going fast enough, any movement into the slipstream felt like you were being hit by a board.

"In my experience, I mostly encountered Zeros, but other guys in our squadron would see the odd Ruff. But for me it was always Zeros. When they attacked, they liked to get on your starboard and port quarter at almost the same altitude and make what I used to call a 'high side run,' but they weren't very high. I can remember specifically in the first battle of the Philippine Sea watching five Zeros on our starboard quarter. They sat just out of range. Of course, we could start firing at a thousand yards, but they were much farther out. These five guys sat there in a loose gaggle when all of a sudden the lead pilot waggled his wings and started his attack run. It was just like in the movies. The second he started banking towards us, I'd started shooting. I don't know if I was hitting him or not, but I think I was.

"We were pretty proud of our three-ship element. If we could get three guys together with a good gunner in the lead plane to step those wingmen up and down, we could suck 'em close and raise a lot of Cain.

"The quality of the Japanese pilots was a lot better at the beginning of the war, but as things went on, they got so bad it was embarrassing. It was embarrassing to me, because here I'm gaining more experience and I hate to say it, but the more I shot back, the more comfortable I become with the idea of someone shooting at me! I had an experience with this Japanese kid who could barely fly his airplane. Here I am rattling the hell out of him with .30- caliber slugs and he'd break off the attack at a considerable distance. They couldn't even fly in formation! When I fought the Japanese during the Battle of Santa Cruz in '42, I was up against their best fighter pilots. They would do loops and play with you. They were just having a ball, but two and a half short years later it was like another world."

Dave Cawley, Dauntless Rear Gunner

Leading Navy SBD Gunners

- ARM2/c John Liska, 4 aircraft confirmed
- ARM2/c WC Colley, 2 aircraft confirmed

Leading Marine Corps SBD Gunners

- Sgt Wallace Read, 3 aircraft confirmed
- Sgt Virgil S Byrd, 2 aircraft confirmed

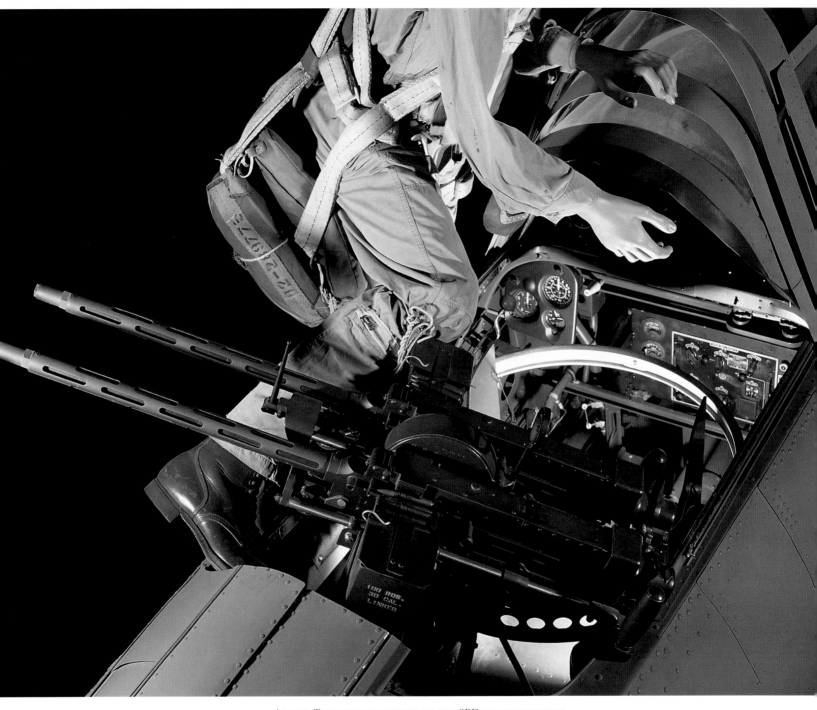

ABOVE: THE GUN MOUNTING ON THE SBD FULLY EXTENDED.
NATIONAL MUSEUM OF NAVAL AVIATION, PENSACOLA.

TOP LEFT: THIS DOUGLAS A-24 (ARMY VERSION OF THE SBD) IS ARMED WITH TWO .30-CALIBER MACHINE GUNS WITH
AMMUNITION BOXES ATTACHED TO THE SIDES. LATER VERSIONS WERE BELT-FED. THE DAUNTLESS WAS EQUIPPED WITH EMERGENCY
FLIGHT CONTROLS FOR THE GUNNERS, SEEN JUST BELOW THE MANNEQUIN'S HAND.
USAF MUSEUM, DAYTON.

GRUMMAN TBF
AVENGER

THE GRUMMAN AVENGER TRULY WAS THE RIGHT AIRCRAFT AT THE RIGHT TIME, AND ONE OF ONLY TWO
CARRIER-BORNE AIRCRAFT EQUIPPED WITH A POWER-DRIVEN TURRET.

In 1941, the Grumman Avenger was the most advanced torpedo bomber in the world. Nicknamed the "Pregnant Beast" and the "Turkey," the portly shaped Avenger proved to be a deadly adversary. Its internal weapons bay and gun turret were a first. Powered by a 1,700-hp Wright Cyclone engine, the Avenger was able to combine impressive range with great bomb-load capacity. But the Avenger's most unique feature was its powered turret. The Grumman-designed turret was the first to be installed on an American carrier-borne aircraft. The first naval aircraft equipped with a powered turret was the British Blackburn Roc.

The Grumman 150SE powered turret was an interesting design. Although almost everything on the Avenger was hydraulic, the turret was electric. The turret perspex was not rounded like other turrets; instead the sides of the turret sloped down to match the angle of the forward canopy. One of the side panels could be removed completely and provided the gunner with an emergency escape. It was an innovative feature not found on most turrets. The turret was also

equipped with a single .50-caliber machine gun instead of the usual two. The single gun was mounted off-center; this gave the gunner an unobstructed field of view in which to aim and acquire his target. The Avenger was also equipped with a single .30-caliber machine gun in the ventral position. This gun was operated by the bomb aimer.

Like most bombers, the Avenger, even with its power turret, was no match against single-seat fighters. In its first combat sortie, six Avengers took off from Midway on June 4, 1942, and headed for the Japanese fleet. The Battle of Midway would go down in history as one of the great victories of World War II, but for the Avenger, it was a disaster. As the unescorted Avengers approached the Japanese fleet, they were intercepted by A6M2 Zeros. It was a one-sided battle. Only one Avenger survived. Severely damaged, it returned to Midway and crash-landed on the beach. It was not a good beginning, but as the war went on, the Avenger proved to be a rugged, dependable torpedo bomber. It was well liked by its crews and would serve well into the 1950s.

ABOVE: AIRCREWMAN DONALD F. FARRELL, AMM2/C, USNR, IN HIS AVENGER TURRET ABOARD THE USS *HORNET* OFF SIAPAN.
RIGHT: A GRUMMAN TBF-1 RUNS UP TO FULL POWER MOMENTS BEFORE TAKE-OFF FROM THE WOODEN DECK OF THE USS *YORKTOWN*.
INSET: THE GRUMMAN TBF-1.

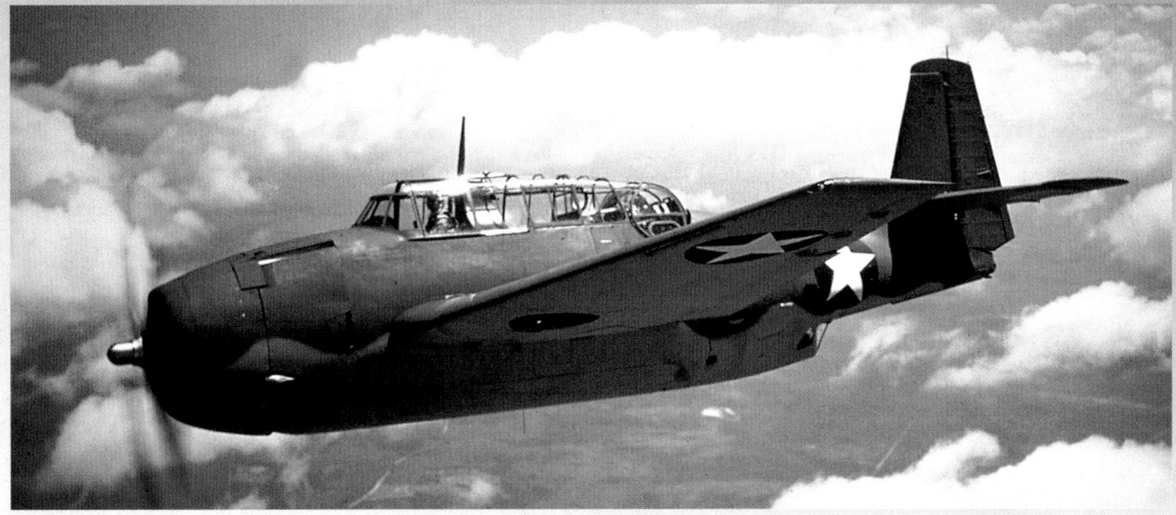

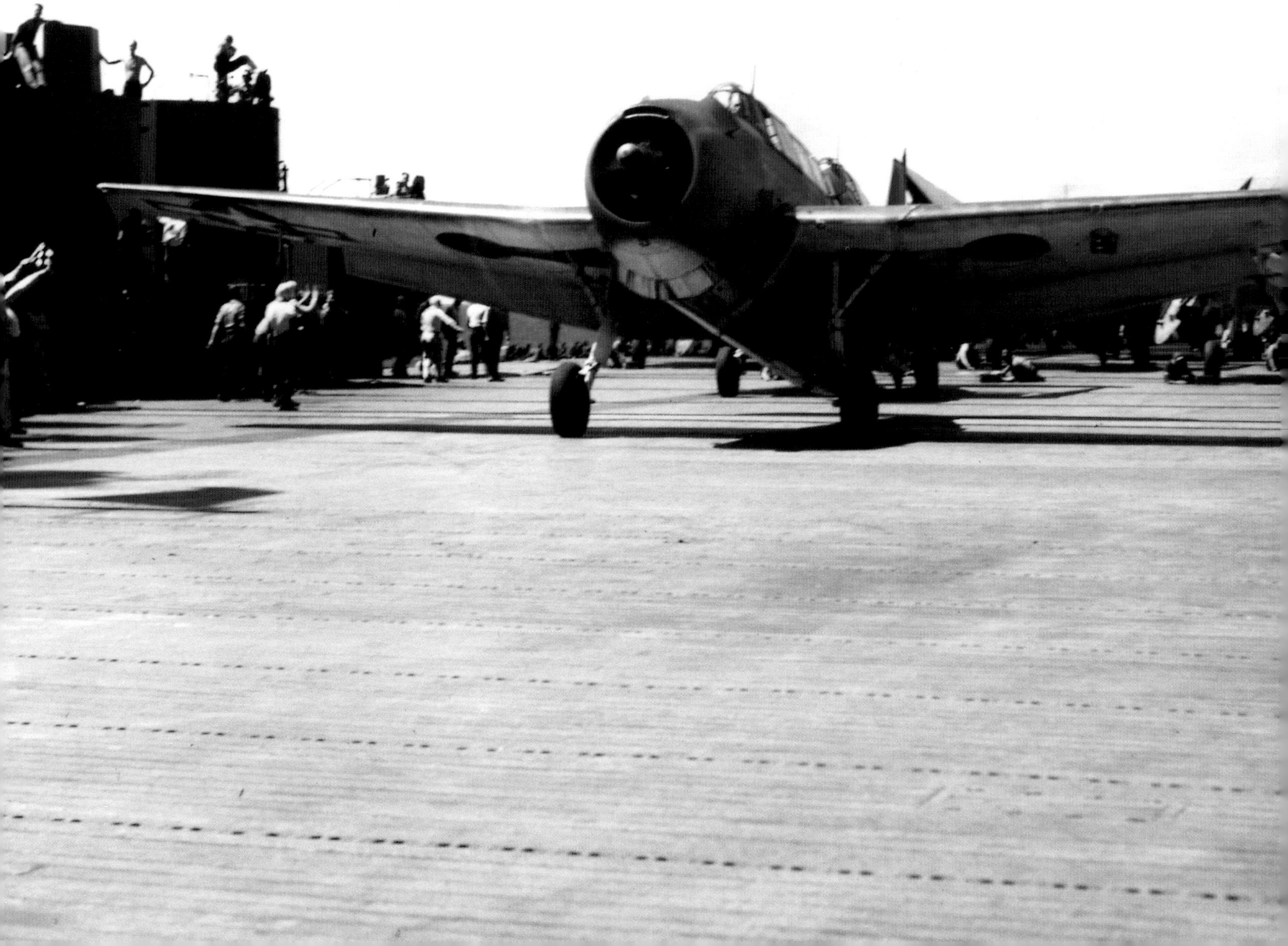

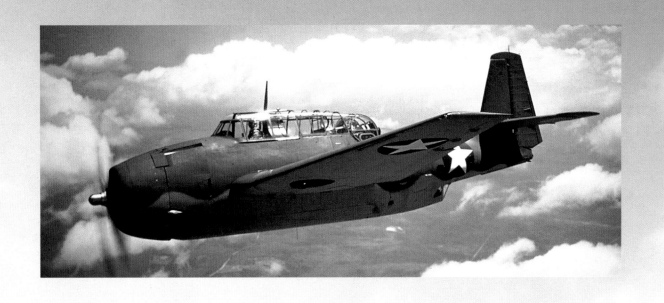

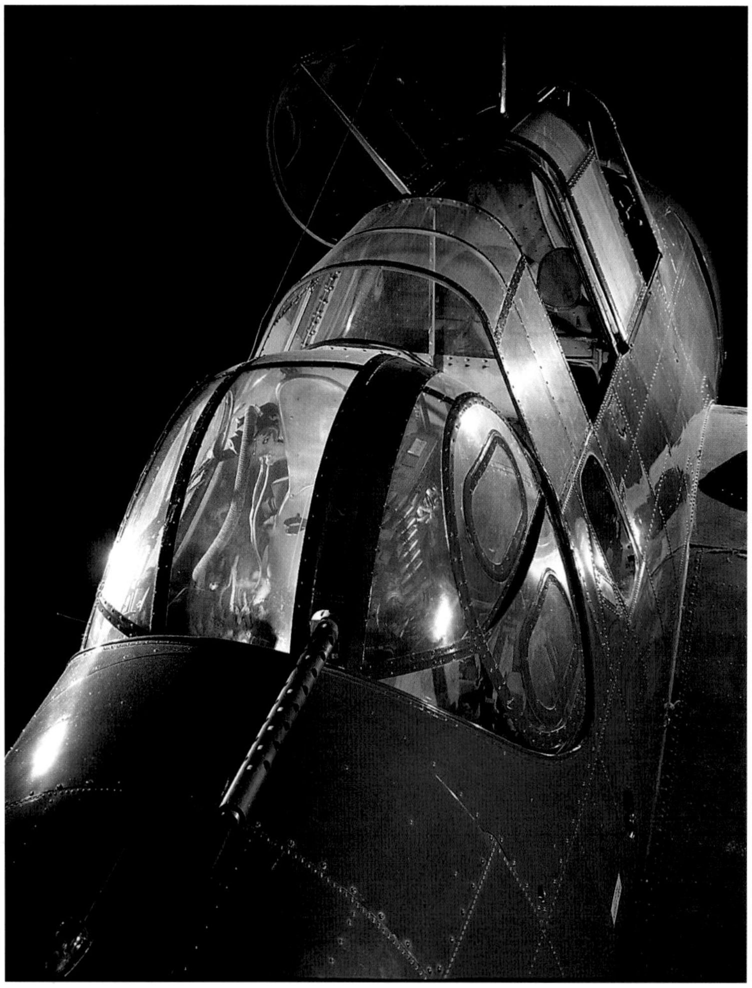

"I served on the USS *Gambier Bay* and USS *Fanshaw Bay*. I was a First Class Aviation Machinist Mate. I had over 500 hours combat flying and made over 150 carrier catapult shots and landings. I joined the Navy as an aviation mechanic. After I came out of mechanic school, I went to go to gunnery school. The first squadron I joined was equipped with the SBD Dauntless dive-bomber. We were stationed in the States at the time, and shortly after, we converted to the TBF Avenger. When we switched to the TBF, it was a whole different world. We went from an open cockpit to an enclosed turret. When we first started training in the Avenger, we had one plane go down into the sea and the gunner had a bit of trouble getting out of the turret's side hatch. It didn't come loose like it should have. After that, we removed all the side hatches from the turret plus the seat belt. In many ways we were still out in the open. We also didn't wear our parachute in the turret. We kept our harness on but not the parachute. There wasn't enough room.

"You sat there pretty balled up. It was a cramped — there wasn't a whole lot of room in those things. On long trips, you could always change off with your radioman. We did that quite a bit. When we were in formation, we had quite a bit of firepower.

"With the Grumman turret, you could swing it so the gun faced forward. From 11 o'clock over to 1 o'clock, you couldn't fire straight forward because of the propeller. But going up to 11 o'clock, I did a lot a shooting. That's where I shot a Japanese plane down. We were coming back from a mission when we saw that our ships were under attack. We spotted this twin-engine aircraft, I believe it was a Fran, low on the water flying away from the ships. My pilot then put the aircraft up on edge, I rotated the turret to the 10 o'clock position and I started firing forward over the wing. I got in a good shot and knocked the Japanese aircraft into the water. The next day we tangled with a Zero.

"We primarily flew ground support mission for the troops. Sometimes I would take four cans of ammunition with me to do some strafing. I would keep maybe fifty, sixty rounds in case we were attacked on the way back to the carrier."

DONALD A. BLANFORD,
FIRST CLASS AVIATION
MACHINIST MATE / GUNNER

ABOVE: THE GRUMMAN TURRET CARRIED A SINGLE .50-CALIBER MACHINE GUN WITH 200 ROUNDS OF AMMUNITION. THIS WAS SUFFICIENT FOR 15 SECONDS OF FIRING, ENOUGH TO COUNTER TWO FIGHTER ATTACKS OR STRAFING OPERATIONS.
NATIONAL MUSEUM OF NAVAL AVIATION, PENSACOLA.

RIGHT: THE INTERIOR OF THE GRUMMAN TURRET. THIS PICTURE CLEARLY SHOWS THE GUNNER'S CONTROL HANDLE, THE 1/4-INCH ARMOR PLATE THAT COVERED THE FRONT OF THE TURRET, AND THE 2-INCH-THICK ARMORED GLASS.
NATIONAL AIR AND SPACE MUSEUM.

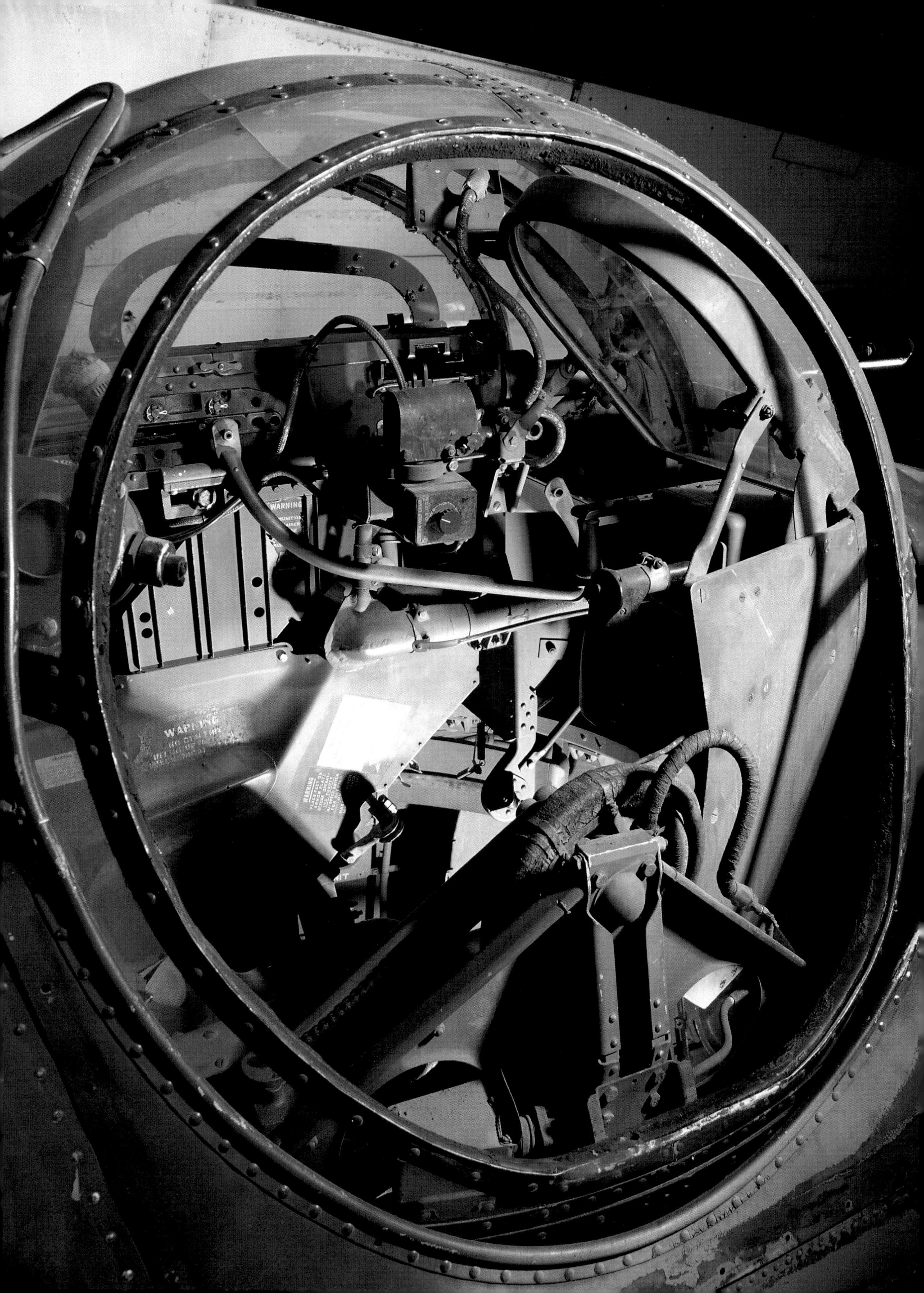
WARNING
WARNING

Fairey Battle

Pitifully armed and slow, the Fairey Battle suffered horrendous losses during the Battle of France and was soon relegated to the training role.

When the Fairey Battle first appeared, it represented a major step forward in combat capability. Compared to the Hawker Hart, which it was designed to replace, the Battle carried twice the bomb load, doubled the range, and was 50 percent faster. It was also the first British aircraft to go into production with the famous Merlin engine.

The Battle's crew consisted of three members, the pilot, bomb aimer / observer and radio operator / gunner. The rear gunner was equipped with a single .303-caliber Vickers K gun. The field of fire from the rear position was limited, and when the rear canopy was open, it was extremely uncomfortable. The Battle's poor performance and weak defensive armament spelled disaster. In 1939, ten squadrons were sent to France as part of the Advanced Air Striking Force. The name may have sounded good in the press, but the force was not advanced and its striking power was woefully inadequate. On May 10, 1940, the

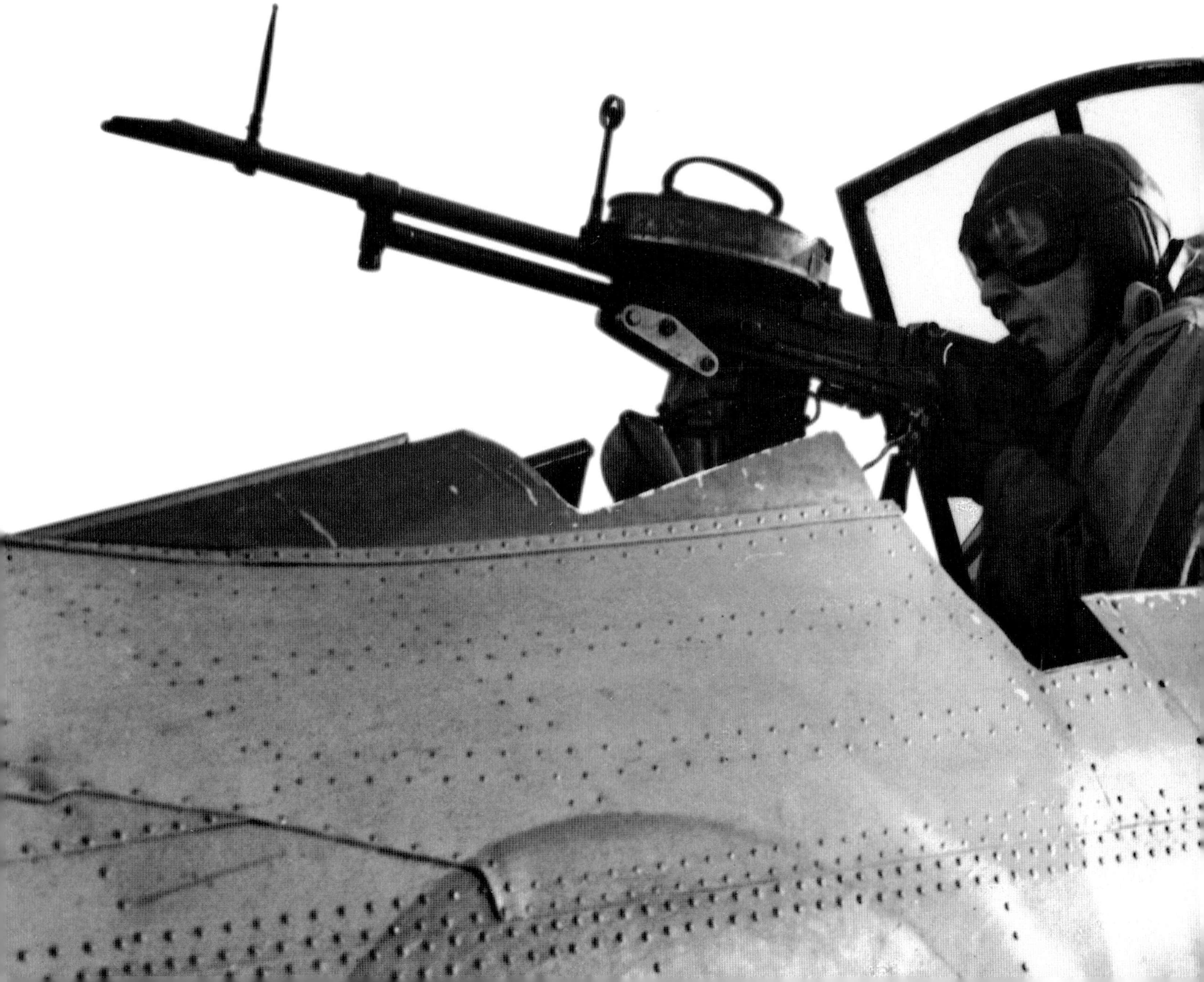

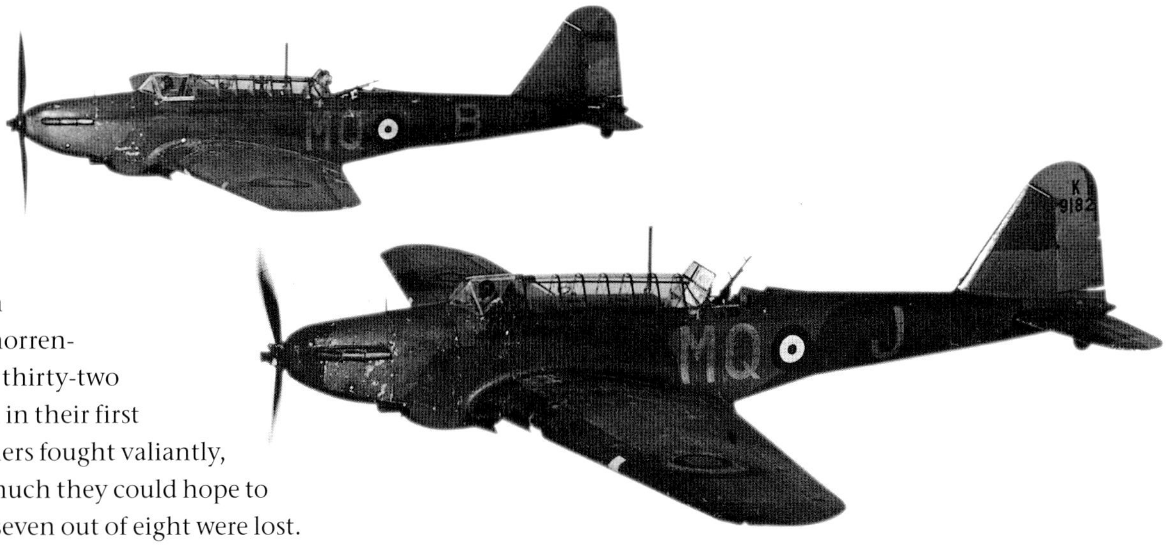
MQ B
K9182
MQ J

invasion of France, Holland and Belgium began. Fairey Battles took to the air and, without fighter escort, attacked the advancing German army. The losses were horrendous — thirteen out of thirty-two Battles were shot down in their first mission. The rear gunners fought valiantly, but there was only so much they could hope to do. The following day seven out of eight were lost.

On May 14, a force of sixty-three Battles were sent to attack troop concentrations and bridges. Thirty-five failed to return. This marked the end of the Battle's day-bomber career. Soon after, the Battle was used as a trainer and for target towing. In the Commonwealth Air Training Plan, the Royal Canadian Air Force used hundreds of Battles to train thousands of air gunners. These men would later serve in the more advanced Lancasters and Halifaxes. Although the Battle's combat career was short, the contribution it made to the Commonwealth Air Training Plan was tremendous. What it lacked in combat performance, the Battle more than made up for it as a trainer.

Below: Armed with the same caliber gun as his World War I counterparts, this Fairy Battle gunner was ill-equipped to deal with the cannon-armed fighters of the Luftwaffe.

Above: A formation of Fairey Battles over France with gunners at the ready.

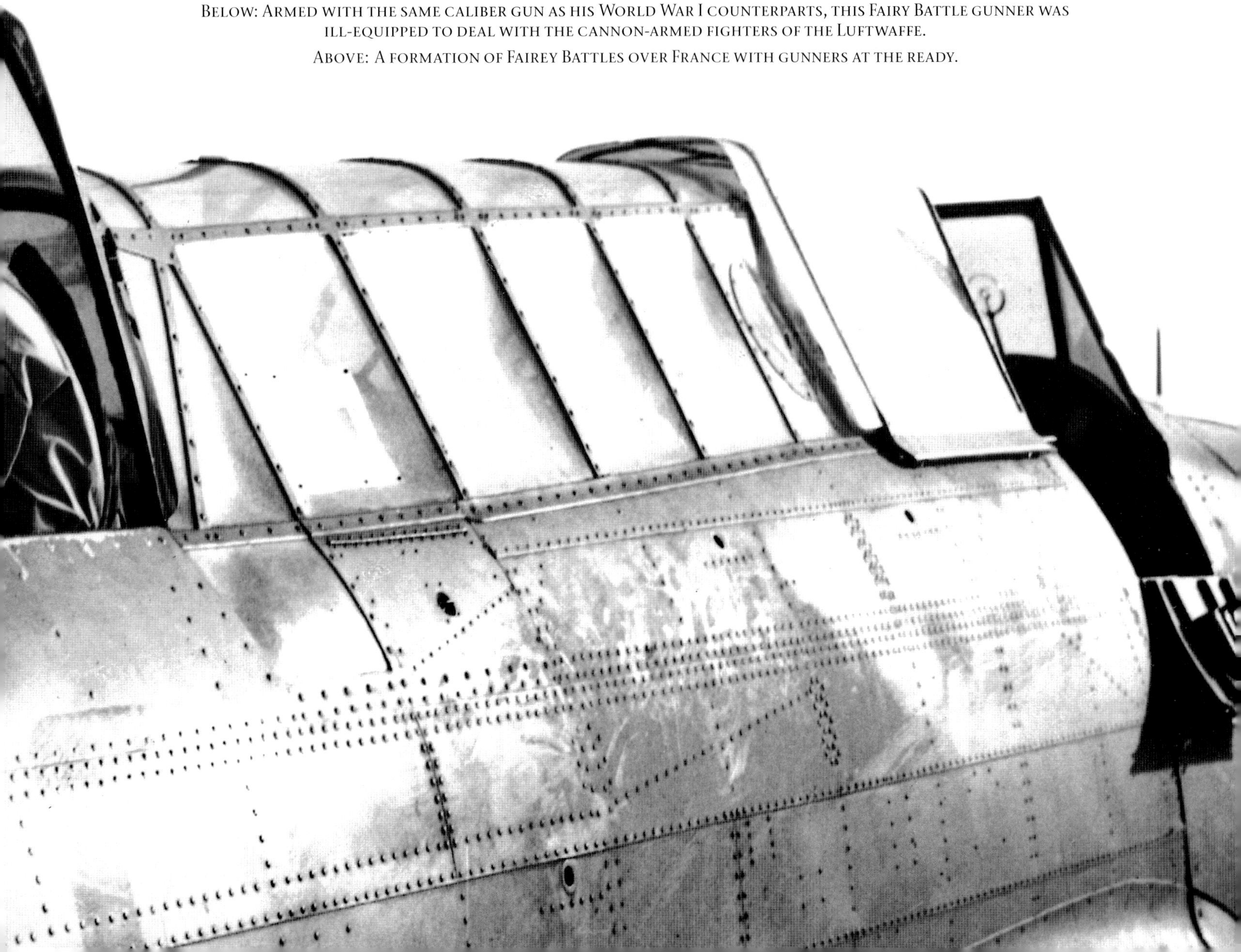

"I went to France with 40 Squadron flying the Fairey Battle. When the Battle was introduced, many great things were said about it, but in fact it was a suicide machine. It was a dreadful thing. In the gunner's position there was peculiar revolving trough that ran longitudinal on the spine of the Battle, located just behind the rear canopy. The gun rested upside down in the trough. That protected it from the elements. When you wanted the gun out you had to revolve it in the trough and pull it up. When the gun was up, it rested on a sort of stalk. It was about 18 inches high and the gun was stuck on top of that. It was just a free gun. There was nothing to stop you from shooting your tail off. Not that anybody ever did. The canopy didn't offer much protection from the slipstream. If you had to open it, you went for it and that's all there was to it. There was also a seat that looked like a piano stool. Of course, it was useless when it came to firing the machine gun. You always had to stand up to do that.

"I don't think anyone, unless you've tried it, can visualize the difficulty of trying to hit a moving airplane from another moving airplane. To start with, the airplane you're in is not very stable and if you can imagine a fighter screaming in on you at 350 miles per hour, the results are not surprising. While you're there, trying to figure out what to do, he's come and gone!"

SERGEANT JOHN HOLMES, AIR OBSERVER

ABOVE: FAIREY BATTLE EQUIPPED WITH A BRISTOL TURRET. AFTER THE BATTLE OF FRANCE, THE BATTLE WAS USED EXTENSIVELY AS A TRAINER. THOUSANDS OF GUNNERS FIRED THEIR FIRST AIR-TO-AIR SHOTS FROM THE FAIREY BATTLE. NATIONAL AVIATION MUSEUM, OTTAWA.

RIGHT: GUNNER'S POSITION IN THE FAIREY BATTLE. RAF MUSEUM, HENDON.

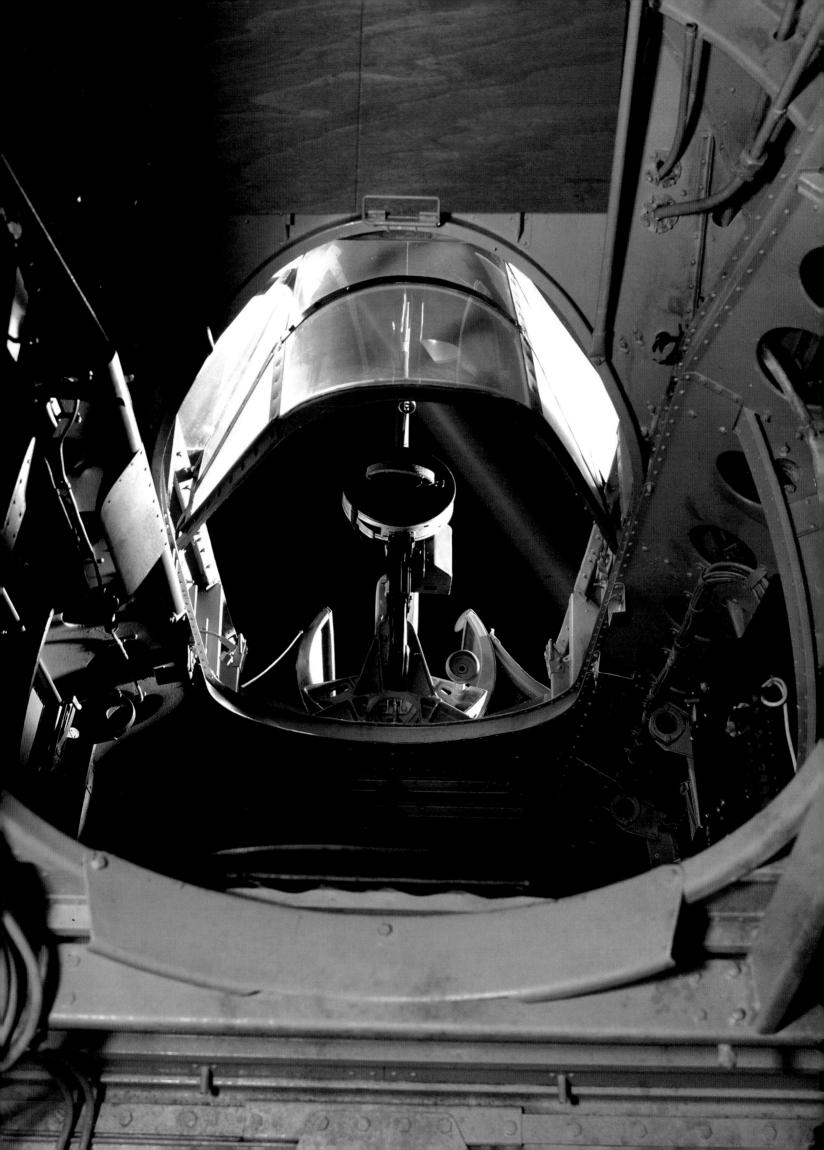

BOULTON PAUL
DEFIANT

AFTER SOME INITIAL SUCCESS AND THEN DRASTIC LOSSES, THE DEFIANT, LIKE THE BOMBERS IN BOMBER
COMMAND, SOUGHT THE COVER OF DARKNESS. THERE IT OPERATED WITH
SOME SUCCESS AS A NIGHT FIGHTER.

When first introduced in the late 1930s, the enclosed powered turret was a marvel of technology and ingenuity — so much so that many believed that bombers armed with such turrets were immune to fighter attack. But what about a fighter armed with such a turret?

In 1935 the British Air Ministry issued Specification F.9/35 calling for a single-engine, two-seater fighter with a powered turret and no wing-mounted machine guns. In 1937 the first prototype, minus the turret, took to the air. Shortly after its first flight, the new Boulton Paul Defiant was fitted with a Boulton Paul Type A Mk IID four-gun turret. The Defiant was in many ways a throwback to the First World War, where the Bristol F2b with a pilot and rear gunner achieved considerable success. The Defiant was also developed in the belief that the gunner, with no responsibility for flying the aircraft and with 360-degree traverse, had more chance of hitting the enemy. Powered by a Rolls-Royce Merlin I, the Defiant had a top speed of 302 mph. Results of comparison trials against the Hawker Hurricane concluded that any average pilot flying a Hurricane could beat the Defiant easily.

In May 1940, the Defiant encountered the Bf 109 for the first time. A flight of Defiants encountered a formation of Ju 87s and quickly dispatched four of the enemy. Unfortunately they were bounced by 109Es and only one of the six turret-armed fighters survived. After that, offensive patrols were canceled and the Defiant was restricted to anti-bomber patrols over the Channel.

During the withdrawal from Dunkirk and the beginning of the Battle of Britain, the Defiant enjoyed a brief moment of glory. Between May 27 and 31, 264 Squadron claimed fifty-seven enemy aircraft destroyed — thirty-seven over Dunkirk on May 29! However, official Lutwaffe records show that only fourteen aircraft were lost in action. The Germans quickly discovered the Defiant's weak spots, and losses mounted at a rapid rate. It was now obvious that the concept of a turret-armed fighter was wrong. The Defiant was withdrawn from day operations and assigned to the night-fighter role.

Given the specifications they were handed, Boulton Paul produced a very good aircraft and turret. It was the idea that was horribly flawed. Although it failed by day, the Defiant proved a formidable night fighter. At its peak, Defiants equipped thirteen night-fighter squadrons.

ABOVE: A DEFIANT NIGHT FIGHTER ON THE RUNWAY AT NIGHT, READY FOR TAKE-OFF.
RIGHT: THIS AIR GUNNER, EQUIPPED WITH THE "RHINO" SUIT, WHICH INCORPORATED A LIFE JACKET AND PARACHUTE,
IS ABOUT TO ENTER HIS TURRET.

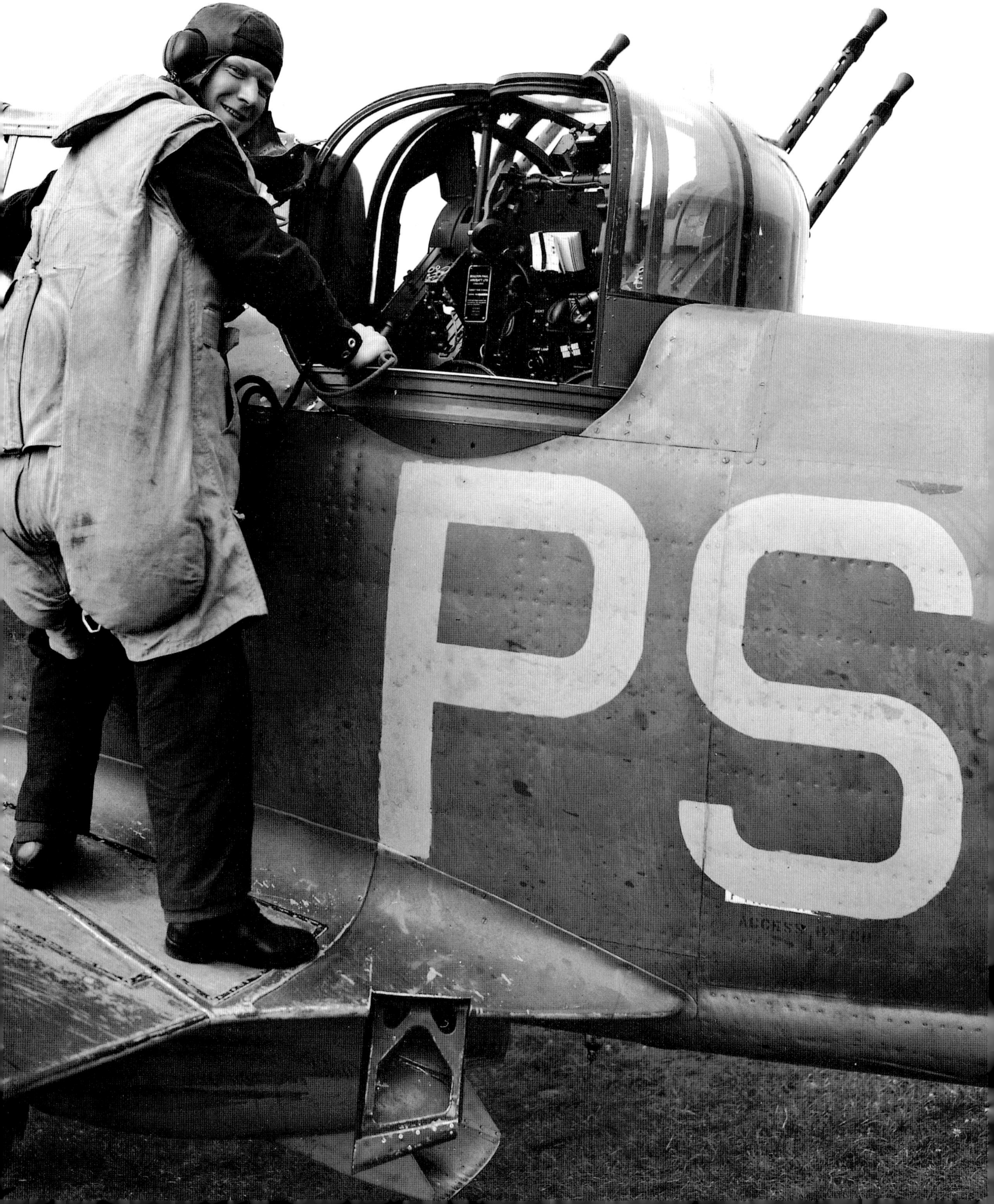
PS

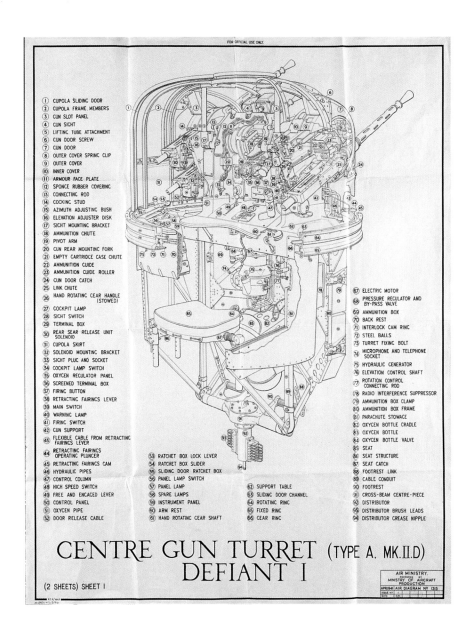

FOR OFFICIAL USE ONLY.

① CUPOLA SLIDING DOOR
② CUPOLA FRAME MEMBERS
③ GUN SLOT PANEL
④ GUN SIGHT
⑤ LIFTING TUBE ATTACHMENT
⑥ GUN DOOR SCREW
⑦ GUN DOOR
⑧ OUTER COVER SPRING CLIP
⑨ OUTER COVER
⑩ INNER COVER
⑪ ARMOUR FACE PLATE
⑫ SPONGE RUBBER COVERING
⑬ CONNECTING ROD
⑭ COCKING STUD
⑮ AZIMUTH ADJUSTING BUSH
⑯ ELEVATION ADJUSTER DISK
⑰ SIGHT MOUNTING BRACKET
⑱ AMMUNITION CHUTE
⑲ PIVOT ARM
⑳ GUN REAR MOUNTING FORK
㉑ EMPTY CARTRIDGE CASE CHUTE
㉒ AMMUNITION GUIDE
㉓ AMMUNITION GUIDE ROLLER
㉔ GUN DOOR CATCH
㉕ LINK CHUTE
㉖ HAND ROTATING GEAR HANDLE (STOWED)
㉗ COCKPIT LAMP
㉘ SIGHT SWITCH
㉙ TERMINAL BOX
㉚ REAR SEAR RELEASE UNIT SOLENOID
㉛ CUPOLA SKIRT
㉜ SOLENOID MOUNTING BRACKET
㉝ SIGHT PLUG AND SOCKET
㉞ COCKPIT LAMP SWITCH
㉟ OXYGEN REGULATOR PANEL
㊱ SCREENED TERMINAL BOX
㊲ FIRING BUTTON
㊳ RETRACTING FAIRINGS LEVER
㊴ MAIN SWITCH
㊵ WARNING LAMP
㊶ FIRING SWITCH
㊷ GUN SUPPORT
㊸ FLEXIBLE CABLE FROM RETRACTING FAIRINGS LEVER
㊹ RETRACTING FAIRINGS OPERATING PLUNGER
㊺ RETRACTING FAIRINGS CAM
㊻ HYDRAULIC PIPES
㊼ CONTROL COLUMN
㊽ HIGH SPEED SWITCH
㊾ FREE AND ENGAGED LEVER
㊿ CONTROL PANEL
51 OXYGEN PIPE
52 DOOR RELEASE CABLE
53 RATCHET BOX LOCK LEVER
54 RATCHET BOX SLIDER
55 SLIDING DOOR RATCHET BOX
56 PANEL LAMP SWITCH
57 PANEL LAMP
58 SPARE LAMPS
59 INSTRUMENT PANEL
60 ARM REST
61 HAND ROTATING GEAR SHAFT
62 SUPPORT TABLE
63 SLIDING DOOR CHANNEL
64 ROTATING RING
65 FIXED RING
66 GEAR RING
67 ELECTRIC MOTOR
68 PRESSURE REGULATOR AND BY-PASS VALVE
69 AMMUNITION BOX
70 BACK REST
71 INTERLOCK CAM RING
72 STEEL BALLS
73 TURRET FIXING BOLT
74 MICROPHONE AND TELEPHONE SOCKET
75 HYDRAULIC GENERATOR
76 ELEVATION CONTROL SHAFT
77 ROTATION CONTROL CONNECTING ROD
78 RADIO INTERFERENCE SUPPRESSOR
79 AMMUNITION BOX CLAMP
80 AMMUNITION BOX FRAME
81 PARACHUTE STOWAGE
82 OXYGEN BOTTLE CRADLE
83 OXYGEN BOTTLE
84 OXYGEN BOTTLE VALVE
85 SEAT
86 SEAT STRUCTURE
87 SEAT CATCH
88 FOOTREST LINK
89 CABLE CONDUIT
90 FOOTREST
91 CROSS-BEAM CENTRE-PIECE
92 DISTRIBUTOR
93 DISTRIBUTOR BRUSH LEADS
94 DISTRIBUTOR GREASE NIPPLE

CENTRE GUN TURRET (TYPE A. MK. II.D) DEFIANT I

(2 SHEETS) SHEET I

AIR MINISTRY.
PREPARED BY
MINISTRY OF AIRCRAFT PRODUCTION
APR 1941 AIR DIAGRAM No 1313

RAF Air Diagram of the Boulton Paul Type A Mk II D type turret.

Opposite: The Type A Mk II D proved to be an efficient design. Its low profile caused minimum drag. Each gun was supplied with 600 rounds of ammunition. RAF Museum, Hendon.

"The Defiant turret. It would be so very easy to complain about the cramped height. Always having one's neck and spine bent into an 'S' shape for the sake of a couple of extra inches. We were only supposed to be five-foot-four anyhow, and I was five-foot-nine. There were not enough short air gunners who were stupid enough to volunteer, so they just took anyone.

"The Rhino suit that we had to wear on Defiants was a bear, but I couldn't come up with an alternative, even though it killed dozens of us. I forgot the details of it, but we could not have sat on our chute or even keep it nearby as in other turrets, so you wore — all in one — an inner layer that fitted a little like a wetsuit of today. The chute fitted around this, and then the dinghy and the outer clothing. There was inner webbing and pockets that literally fell apart (I presume) when one bailed out. Getting into the turret was awkward, to say the least. Sliding on your back through the tiny section of a door — like a partly skinned orange — then half in, you had to grope backward and down for the hanging webbing seat, which was really just a wide strap. Somehow you hooked it up and tried to get comfortable. You would bang your head or arms on everything.

"Once in I felt like a god. A mere air gunner who got to fly fighters. To be thrown around the sky behind a Merlin was a great thrill. But if getting in was hard, getting out while flying or crashing must have been impossible. I can't remember anyone who made it, so what use was the Rhino suit? Maybe it gave us some sort of courage, even if we knew that the odds of it saving our lives was minuscule."

FREDERICK "GUS" PLATTS,
AIR GUNNER,
230, 282 AND 280 SQUADRONS

150

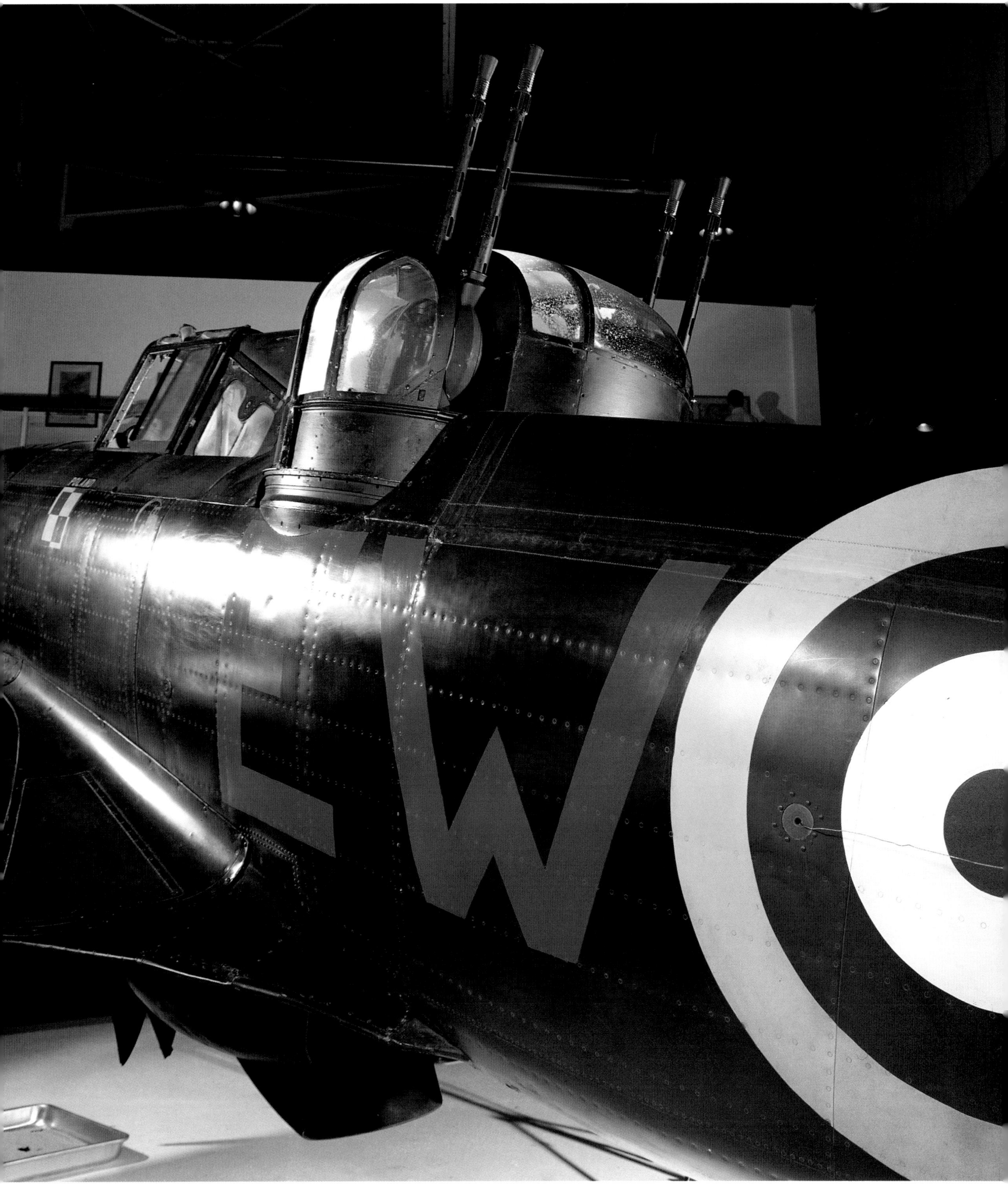

ILYUSHIN IL-2
SHTURMOVIK

HEAVILY ARMED AND ARMORED, THE SHTURMOVIK WAS ONE OF THE MOST FEARED GROUND-ATTACK
AIRCRAFT OF THE SECOND WORLD WAR.

The Il-2 Shturmovik was produced in larger numbers than any single type of aircraft in history. It was also one of the most effective ground-attack aircraft ever conceived. During most of the Second World War, the USSR pumped out an average of 1,200 aircraft per month for a grand total of 35,952 Il-2s.

The Il-2 was the Russian Air Force's most effective and celebrated aircraft. The West knew it as the Shturmovik, but for the combatants on the Eastern front it was called *Ilyusha* by the Russians and *Schwarz Tod* (Black Death) by the Germans.

In the 1930s, the Red Air Force saw their aviation assets as a tactical force, a force that would support the army exclusively. This called for a self-sufficient, fast and rugged ground-attack aircraft capable of destroying tanks, fortifications and infantry. Several designs were submitted, but it was the CKB-37, designed by Sergei V. Ilyushin, that won the day. Shortly after, the Il-2 was ordered into production.

The original Il-2 was a single-seater. Early encounters with German fighters showed the Il-2 to be extremely vulnerable. In 1941 the Soviet High Command was reluctant

IL-2M IN FLIGHT. WHILE THE COCKPIT AND FRONT OF THE IL-2 WAS PROTECTED BY AN ARMOR PLATING
5 TO 12 MM THICK, THE GUNNER HAD LITTLE OR NO ARMOR PLATE FOR PROTECTION.

to admit the high attrition rate suffered by the Il-2. The Il-2 also lacked sufficient punch to defeat enemy armor. A redesign was ordered. The impressive armored area that covered the pilot's cockpit (between 5 and 12 mm thick) was extended to accommodate a rear gunner. A heavy 12.7-mm BS or Berezin UBT machine gun was fitted to cover the upper rear area, and two high-velocity 23-mm cannon replaced the original 20-mm wing-mounted cannons. The new two-seater entered service in August 1942. Some success was achieved against German fighter pilots who assumed they were attacking the regular single-seat version. This did not last long and losses again began to mount. Even though it was heavily armed and armored, the Il-2 suffered severe losses.

In 1943 the Soviet Air Force achieved numerical superiority over the Luftwaffe, but not air superiority over the battlefield. The respect for the Luftwaffe was so great that one-for-one fighter escort was provided for the Il-2. In 1943 the Luftwaffe claimed 6,900 Il-2s shot down, with a good many shot down by new armored mobile anti-aircraft guns. At the beginning of the Battle of Kursk, the Il-2 was using the new 132-mm rocket. These projectiles carried a hollow charge warhead and proved absolutely devastating when used against fixed emplacements. The Il-2 operated in the extreme low-level regime. This is where it was most effective, so many missions were flown below 200 feet. At this level, surprise was often complete. The favorite form of attack was to approach the target from the rear. This often assured surprise, and gave the Il-2s a better chance of avoiding

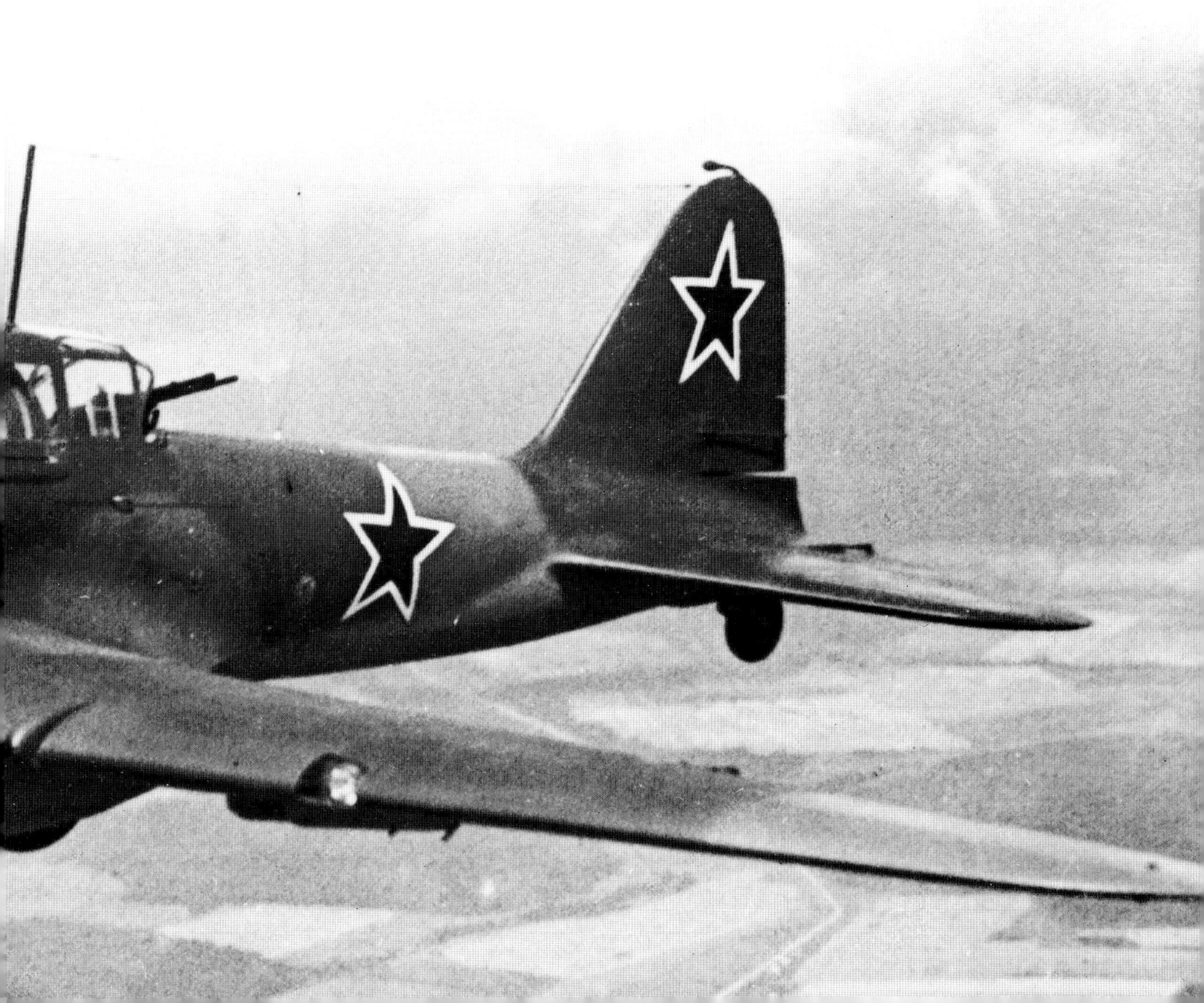

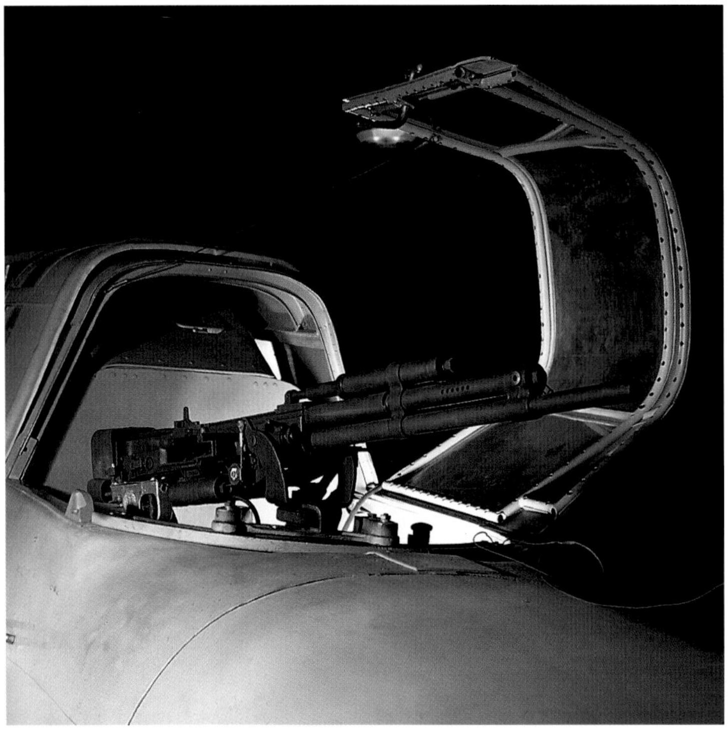

THE NATIONAL AIR AND SPACE MUSEUM'S IL-2M3. GUNNERS IN THE IL-2 WERE EXTREMELY VULNERABLE TO ENEMY FIGHTERS AND WERE KILLED SEVEN TIMES MORE OFTEN THAN THEIR PILOTS.

RIGHT: CREATURE COMFORTS FOR THE REAR GUNNER WERE NON-EXISTENT, AS CLEARLY SHOWN BY THE CANVAS STRAP FOR A SEAT. THE ADDITION OF A REAR GUNNER WITH A HEAVY CALIBER 12.7-MM BS MACHINE GUN IMPROVED THE IL-2'S REAR DEFENSE TO A DEGREE, BUT ONCE LUFTWAFFE PILOTS ADJUSTED THEIR TACTICS, THE IL-2 CONTINUED TO SUFFER HEAVY LOSSES.
NATIONAL AIR AND SPACE MUSEUM.

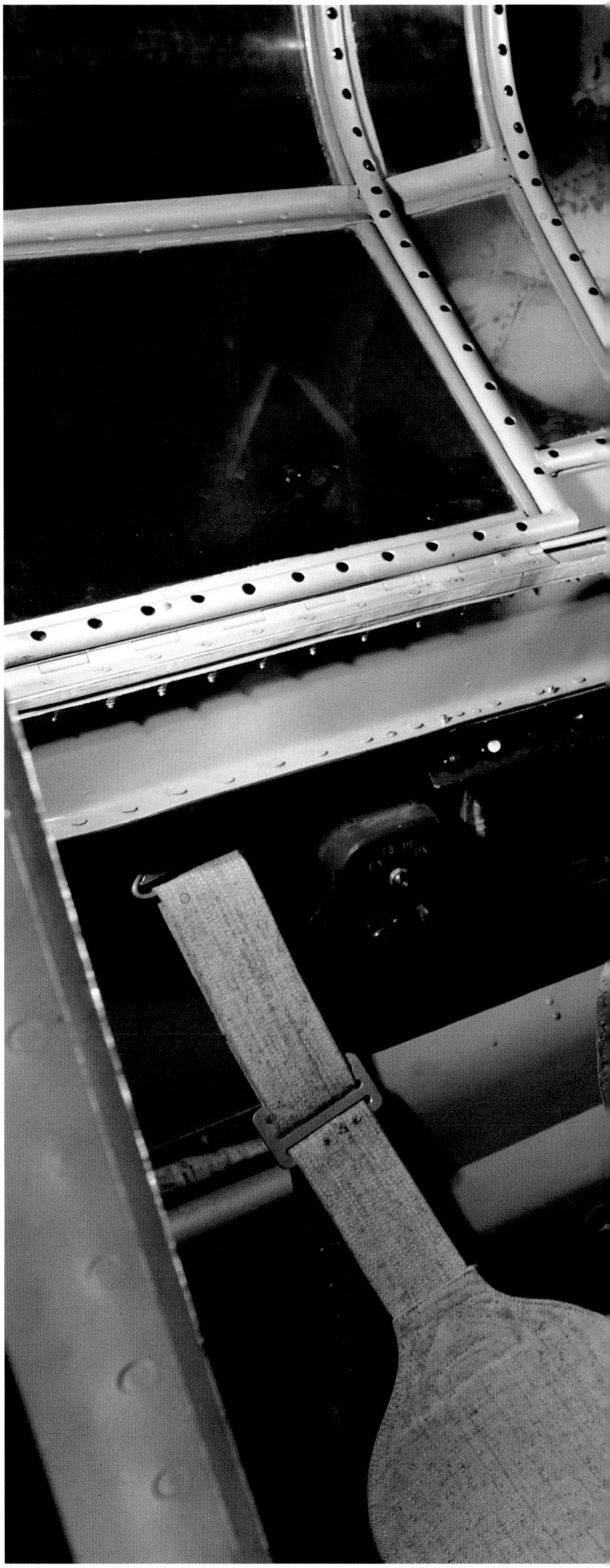

the anti-aircraft defenses and a higher probability of destroyed tanks. This form of attack, with some minor modifications, remained until the end of the war.

For all its losses, the Il-2 proved a very effective ground-attack aircraft. During the gigantic battle of Kursk, according to Soviet records, Il-2s attacked the German 9th, 3rd and 17th Panzer divisions and scored a major victory. In just over 20 minutes, Il-2 Shturmoviks destroyed 70 German tanks in the 9th Panzer Division; 270 were lost in the 3rd Panzer Division in just over 120 minutes; and in an attack that lasted 240 minutes, the 17th Panzer Division lost 240 tanks!

By late 1943 the Soviet Air Force was better trained and equipped. Aircraft that had started the war had been replaced with newer versions that were as good or better than their German counterparts. Their combat techniques had improved as well, but the cost had been enormous. Until the very end, T-34 tanks supported by regiments of Shturmoviks ravaged German tank and infantry formations. No matter how many tanks or Il-2s the Germans destroyed, new regiments would appear on the battlefield to continue the fight. In February 1945 the final version of the Shturmovik to fly in World War II appeared over the Eastern Front. The Il-10 was more streamlined and powered by a strong 2,000-hp engine.

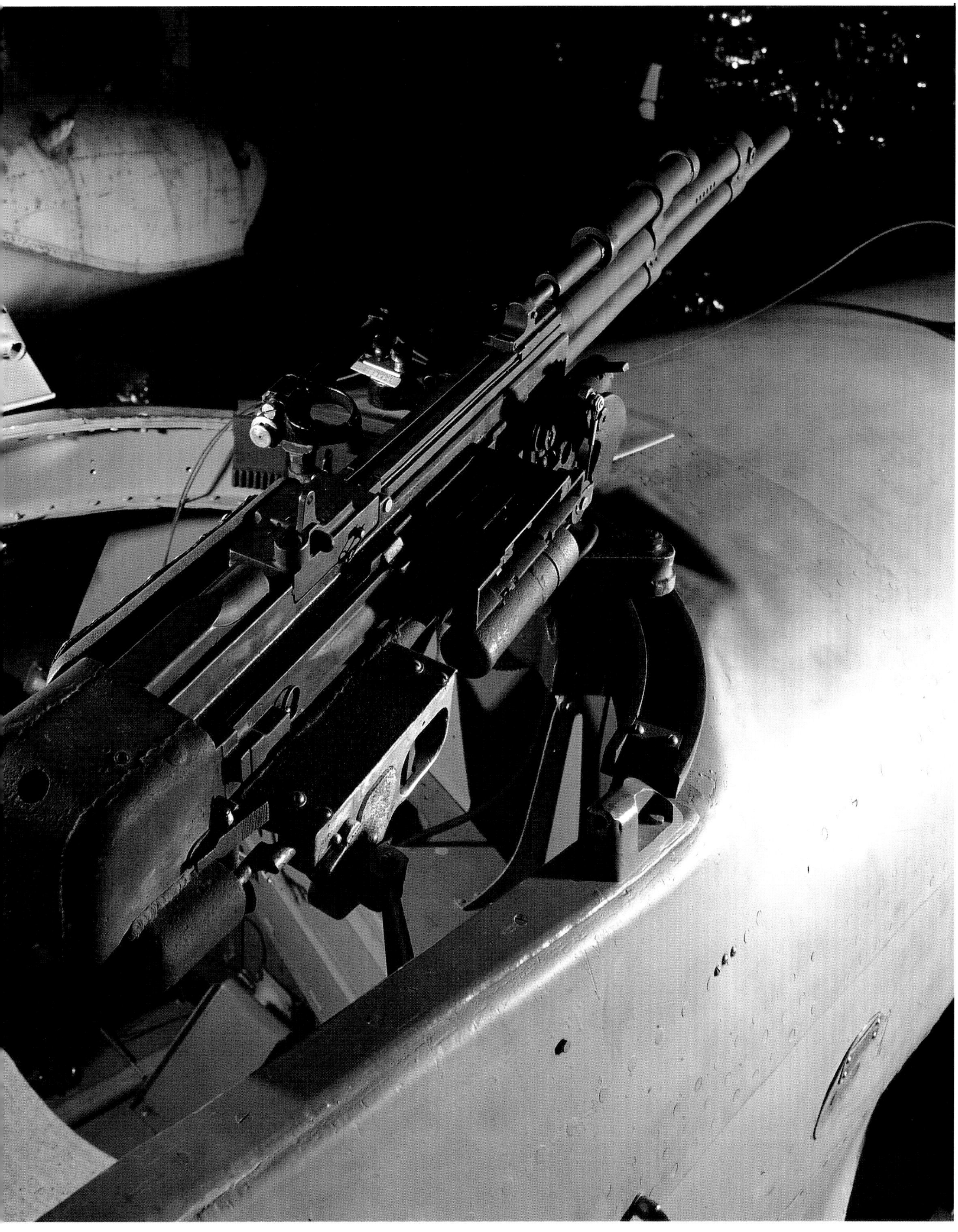

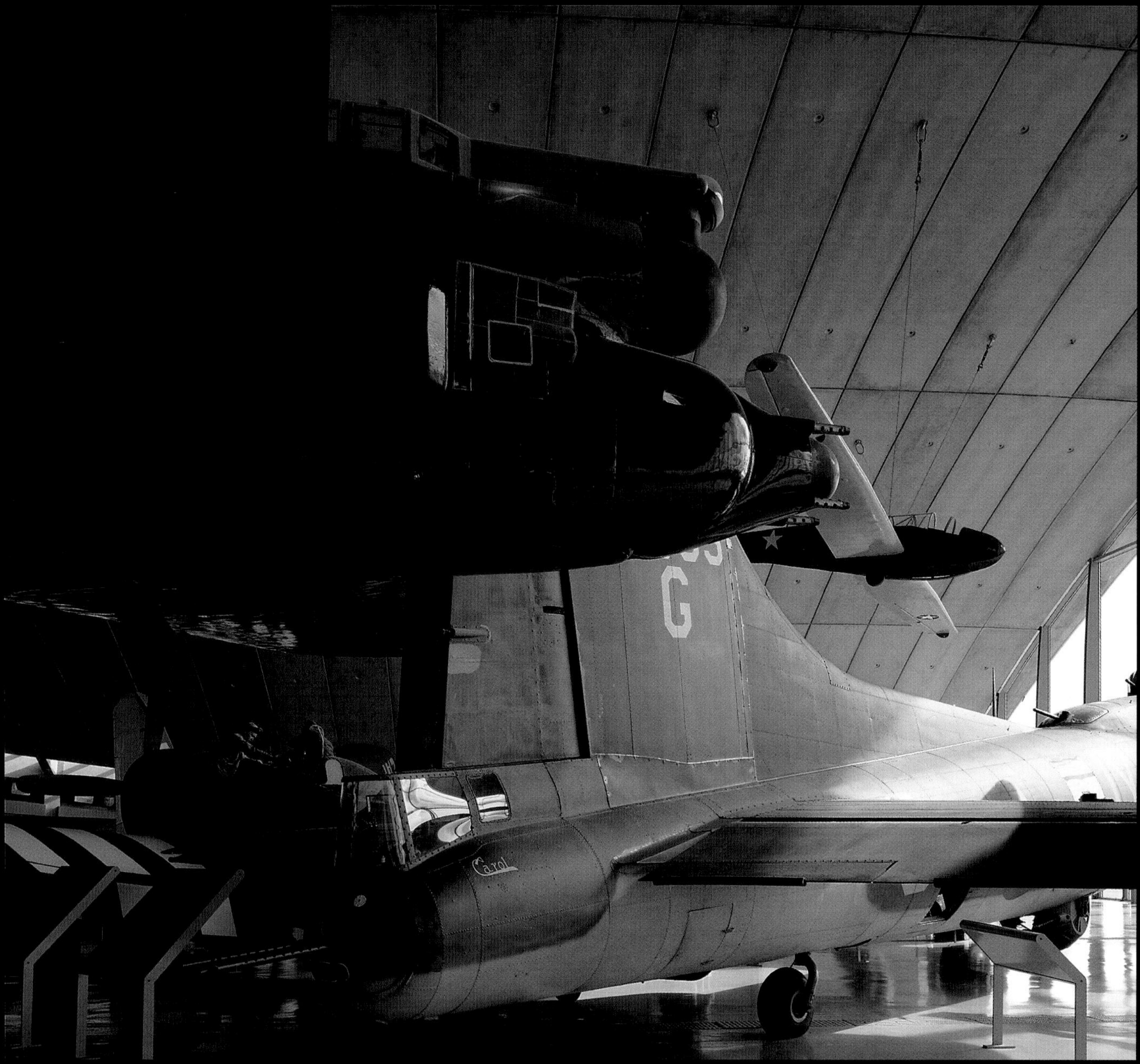
G
Carol

At the end of the Second World War, the air gunner all but disappeared. How effective he was is hard to measure in pure numbers alone. The USAF History Office has not recognized claims from gunners due to the obvious problems of confirmation since there would be more than one gunner firing at the same enemy fighter, but records were kept. According to 8th Air Force Bomber Operations records, air gunners, under "allowed combat claims," shot down 10,326 German fighters between August 1942 and September 1944. Postwar research showed that these claims amounted to a small fraction of what the Luftwaffe actually lost. What is clear is that the air gunner's role in the destruction of the Luftwaffe and Japanese Air Forces was considerable. Without their vigilance and protection, many USAAF and RAF crews would not have survived. During the Korean War, B-29 gunners were credited with sixteen MiG 15s and two decades later, two B-52 tail gunners shot down two MiG 21s over North Vietnam. Three other MiGs were claimed but not confirmed. The air gunner's role in the history of 20th-century warfare was unique and lasted just fifty-eight years.

The Cheyenne tail position on a Boeing B-17G. Above is the four-gun tail position of the Boeing B-52 nuclear bomber. Both are armed with .50-caliber machine guns.
American Air Power Museum, Duxford

ACKNOWLEDGMENTS

This book would not have been possible without the help, support and encouragement of my family — my parents, Johan and Tonia, my brother, Gordon, and his wife, Lydia, my sister, Marilyn, my nephew, Eric, and my two nieces, Julia and Andrea. I would like to thank John Denison and everyone at the Boston Mills Press and Stoddart Publishing. To Victoria Ward for her friendship, inspiration and insights. To Dan Patterson, for his dedication and friendship — you the man, Dan. What should we do next?

The following people gave their time and expertise to help make this book possible. Grateful thanks to Scott Thompson, Daniela Tombari, Donald Bouchad, Dave Menard, Gary Sanford, Robert Linnell, Fiona Hale, Steve Payne, John T. Moyles of the Air Gunners' Association of Canada, Chuck Hansen, Suzanne Depoe, Gunther Rall, Frederick Gus Platts, Maureen Bracegirdle, Peter Spoden, Al Parisani, Fred Vincent, Bill Cole, Joe Davis, Tony Holmes, Dennis David, Ron Dick, John Holmes, Peter Caldwell, Peter Underhill, Barrett Tillman, David J. Cawley, Douglas L. Sample, Jim Rodella, Bill Agee, Robert Blumberg, Brian Ron Bramley, Philip Jarrett, Eric Cameron, John Hutton, Joe Taddonio, Rene St. Pierre, Joe Ouellette, Marvin Yost, Andy Doty, Donald A. Blanford, Karl Johanssen, Contessa Maria Fede Armani Caproni, Piera Graffer, Diana Bachert, Teresa Lacy, David Tallichet, Clive and Linda Denney, Tony Ditheridge, Bob Ready, Squadron Leader Paul Day BBMF/RAF, Tom Alison, and Tom Patterson. The design for this work was inspired by the talented work of Kevan Buss. Thanks, buddy.

I would also like to thank the following museums: the National Aviation Museum, Ottawa; the RCAF Museum, Trenton; the USAF Museum, Dayton; the RAF Museum, Hendon; the Aerospace Museum, Cosford; the Aircraft Restoration Company, Duxford; the National Museum of Naval Aviation, Pensacola; the National Air and Space Museum, Washington; the Museo Storico del' Aeronautica Militaire, Rome; the Imperial War Museum, Duxford.

DONALD NIJBOER

PHOTOGRAPH CREDITS

(I) *Inset* (M) *Main* (T) *Top* (B) *Bottom*

390th Bomb Group Collection: 28(B), 44(M), 51(I, B)
Author's Collection: 11, 16, 38(T), 53(I), 62(I), 89, 128(T), 137
Boeing: 67
Bundesarchiv: 30, 119(M, I)
Canadian National Archives: 32, 72, 110, 144(M)
Canadian War Museum: 123
Cheryl Behner Patterson: 9
Chuck Hansen: 47(B), 50(B), 63(T), 88, 100
F. Bradly Peyton III: 57(I), 63(B)
Imperial War Museum: 7, 8, 13, 19, 22, 25(T), 31(B), 76, 80, 81(I), 82, 90(T), 99(I), 103(I), 106, 107, 114, 115, 116(B), 118, 120, 122(M), 124(T, B), 130, 131, 145(I), 148, 149
Jeffrey L. Ethell Collection: 27, 28(T), 29(T), 45(I), 56(M), 66(M, I), 92, 98(M), 102(M), 141(I)
John Weal:116(T)
National Air & Space Museum: 29(B), 126(M), 127
National Archives, D.C.: 3, 5, 18(B), 20, 21(T, B), 25(B), 26, 33, 34(T), 35, 39(B), 93, 102(I), 130(T), 136, 140, 141(M)
Philip Jarrett: 31(T), 34(B), 133, 152
RAF Museum: 75, 78, 81(M), 84, 111, 121, 150
Tony Ditheridge: 90(B)
USAFM: 60(T), 94(B), 96(T, B)

Page 42
Top left, Jeffrey L. Ethell Collection; *top right,* National Archives, D.C.; *bottom right,* Jeffrey L. Ethell Collection
Page 43
Left and right, RAF Museum

Page 86
Left, top to bottom: Jeffrey L. Ethell Collection; National Archives, D.C.; Imperial War Museum
Right, top to bottom: Jeffrey L. Ethell Collection; Jeffrey L. Ethell Collection; RAF Museum
Page 87
Left, top to bottom: Imperial War Museum; Philip Jarrett
Right, top to bottom: National Air and Space Museum; National Air and Space Museum; Philip Jarrett

Page 134
Left to right: National Archives, D.C.; Jeffrey L. Ethell Collection
Page 135
Top left: Canadian National Archives; *top right,* Imperial War Museum; *bottom right,* Philip Jarrett

All other photographs by Dan Patterson

BIBLIOGRAPHY

Anderton, David A. *Aggressors Volume 3: Interceptor versus Heavy Bomber.* Tokyo/New York: Zokeisha Publications Ltd., 1991

Bowan, Ezra. *Knights of the Air.* Chicago, Illinois: Time-Life Books, 1980

Bowan, Martin W. U*SAF Handbook 1939–1945.* Mechanicsburg, PA: Stackpole Books, 1997

Boyer, Chaz. *Guns in the Sky: The Air Gunners of World War II.* London: J. M. Dent & Sons Ltd., 1979

Bruchiss, Louis. *Aircraft Armament.* New York, NY: Aerosphere, Inc., 1945

Clark, R. Wallace. *British Aircraft Armament, Volume 1: RAF Gun Turrets from 1914 to the Present Day.* Sparkford Nr Yeovil, Somerset: Patrick Stephens Limited, 1993

Dick, Ron, Dan Patterson. *American Eagles: A History of the United States Air Force.* Charlottesville, VA: Howell Press, 1997

Doty, Andy. *Backwards into Battle: A Tail Gunner's Journey in World War II.* Palo Alto, California: Tall Tree Press of Palo Alto, 1995

Donald, David. *American Warplanes of World War II.* London, England: Aerospace Publishing Ltd., 1995

———. *Warplanes of the Luftwaffe.* London, England: Aerospace Publishing Ltd., 1995

Dorr, Robert F. *B-24 Liberator Units of the Eighth Air Force.* Botley, Oxford: Osprey Publishing, 1999

Everitt, Chris, Martin Middlebrook. *The Bomber Command War Diaries: An Operational Reference Book: 1939–1945.* London, England: Penguin Books Ltd., 1990

Frankland, Noble. *Bomber Offensive: The Devastation of Europe.* New York, NY: Ballantine Books Inc., 1970

Franks, Norman. *Aircraft Versus Aircraft: The Illustrated History of Fighter Pilot Combat since 1914.* New York: Cresent Books, 1986

Green, William. *Famous Bombers of the Second World War.* 16 Maddox Street, W.1.: Macdonald & Co. (Publishers) Ltd., 1959

Greenhouse, Brereton, Stephen J. Harris, William C. Johnston, William G. P. Rawling. *The Crucible of War 1939–1945: The Official History of the Royal Canadian Air Force Volume III.* Toronto: University of Toronto Press Inc. in cooperation with the Department of National Defence and the Canada Communication Group, Publishing, Supply Services of Canada, 1994

Hansen, Chuck. *American Aircraft Turrets and Flexible Gun Mounts 1917–1960.* Unpublished, 1976

Hastings, Max. *Bomber Command: Churchill's Epic Campaign.* New York, NY: Touchstone, Simon & Schuster, 1989

Hess, William. *B-17 Flying Fortress.* New York, NY: Ballantine Books Inc., 1974

Hoffschmidt, *Edward J. German Aircraft Guns WWI–WWII.* Old Greenwich, Conn.: WE Inc., 1969

Jablonski, Edward. *Double Strike: The Epic Air Raids on Regensburg/Schweinfurt.* Garden City, NY: Doubleday Company, Inc., 1974

Lake, Jon. *Halifax Squadrons of World War 2.* Botley, Oxford: Osprey Publishing, 1999

Levine, Alan J. *The Strategic Bombing of Germany.* Westport, CT: Praeger Publishers, 1992

March, Daniel J. ed. *British Warplanes of World War II.* London, England: Aerospace Publishing Ltd., 1998

Mesko, Jim. *A-20 Havoc in Action.* Carrollton, TX: Squadron/Signal Publications, 1994

Middlebrook, Martin. *The Nuremberg Raid.* Middlesex, England: Penguin Books Ltd., 1986

———. *The Berlin Raids: RAF Bomber Command Winter 1943–44.* Middlesex, England: Penguin Books Ltd., 1988

Patterson, Dan, Paul Perkins. *The Lady: Boeing B-17 Flying Fortress.* Charlottesville, VA: Howell Press Inc., 1993

Price, Alfred. *The Bomber in World War II.* London: Macdonald and Jane's, 1979

Punka, George. *Messerschmitt Me 210/410 in Action.* Carrollton, TX: Squadron / Signal Publications, 1994

Sweetman, John. *Schweinfurt Disaster in the Skies.* New York, NY: Ballantine Books Inc., 1971

Taylor, John W. R. *A History of Aerial Warfare.* Middlesex, England: The Hamlyn Publishing Group Limited, 1974

Tillman, Barrett. *SBD Dauntless Units of World War 2.* Botley, Oxford: Osprey Publishing, 1998

Vanags-Baginkskis, Alex. *Aggressors, Volume 1: Tank Buster versus Combat Vehicle.* Tokyo/New York: Zokeisha Publications Ltd., 1990

Wallace, G. F. *The Guns of the Royal Air Force 1939–1945.* London: William Kimber and Co. Limited, 1972

Weal, John. *Messerschmitt Bf 110 Zerstörer Aces of World War 2.* Botley, Oxford: Osprey Publishing, 1999

Wood, Alan C., Terry C. Treadwell. *The First Air War: A Pictorial History 1914–1919.* London: Brassey's, 1996

Periodicals

Allward, Maurice. "Aircraft Gun Turrets Prior to the Second World War." *The Putnam Aeronautical Review.* March 1990

Brown, J. "Remote-Controlled Turret in the Boeing B-29." *Air Enthusiast Quarterly.* Journal No. 3

Bushby, J. R. "British Bomber Defence, 1939–1945." *The Putnam Aeronautical Review.* September 1990

Hansen, Chuck. "Flying Guns: Shooting up the Stratosphere...The First Quarter Century Airpower." *Wings.* August 1980

————. "Grumman Gun Turret." *American Aviation Historical Society Journal,* Vol. 32, No. 3. Fall 1987

Public Records

The Public Record Office, Kew Gardens, London.

AIR 2, AIR 14, AIR 16, AIR 46

The RAF Museum, Hendon

Special Informational Intelligence Report No. 43–17. German Fighter Tactics Against Flying Fortresses. RAF Museum

The Imperial War Museum, London

SchuBwaffenanlage BedienuwgsuorschriPt-Wa Flugzeug-Handbuch He 111H-16, Ju 88A-4 und D-1 and Bf 110G-4(N)

The National Archives, Washington

Headquarters, Eighth Air Force Operational Analysis Section. *An Evaluation of Defensive Measures Taken to Protect Heavy Bombers from Loss and Damage.* November 1944

Combat Analysis Study Number 5 and an Evaluation by Intelligence. Japanese Aircraft vs U.S. Aircraft in Aerial Combat, Southwest Pacific Theater, June 5, 1943

The National Aviation Museum, Ottawa

USAF Historical Study No. 136: Development of the Long-Range Escort Fighter. USAF Historical Division Research Studies Institute Air University, September 1955

INDEX